Lecture Notes in Electrical Engineering

Volume 623

The book series *Lecture Notes in Electrical Engineering* (LNEE) publishes the latest developments in Electrical Engineering—quickly, informally and in high quality. While original research reported in proceedings and monographs has traditionally formed the core of LNEE, we also encourage authors to submit books devoted to supporting student education and professional training in the various fields and applications areas of electrical engineering. The series cover classical and emerging topics concerning:

- Communication Engineering, Information Theory and Networks
- Electronics Engineering and Microelectronics
- Signal, Image and Speech Processing
- Wireless and Mobile Communication
- Circuits and Systems
- Energy Systems, Power Electronics and Electrical Machines
- Electro-optical Engineering
- Instrumentation Engineering
- Avionics Engineering
- Control Systems
- Internet-of-Things and Cybersecurity
- Biomedical Devices, MEMS and NEMS

For general information about this book series, comments or suggestions, please contact leontina.dicecco@springer.com.

To submit a proposal or request further information, please contact the Publishing Editor in your country:

China

Jasmine Dou, Associate Editor (jasmine.dou@springer.com)

India, Japan, Rest of Asia

Swati Meherishi, Executive Editor (Swati.Meherishi@springer.com)

Southeast Asia, Australia, New Zealand

Ramesh Nath Premnath, Editor (ramesh.premnath@springernature.com)

USA, Canada:

Michael Luby, Senior Editor (michael.luby@springer.com)

All other Countries:

Leontina Di Cecco, Senior Editor (leontina.dicecco@springer.com)

** Indexing: The books of this series are submitted to ISI Proceedings, EI-Compendex, SCOPUS, MetaPress, Web of Science and Springerlink **

More information about this series at http://www.springer.com/series/7818

Dhanasekharan Natarajan

Fundamentals of Digital
Electronics

 Springer

Dhanasekharan Natarajan
Bengaluru, India

ISSN 1876-1100   ISSN 1876-1119   (electronic)
Lecture Notes in Electrical Engineering
ISBN 978-3-030-36198-3   ISBN 978-3-030-36196-9   (eBook)
https://doi.org/10.1007/978-3-030-36196-9

This Springer imprint is published by the registered company Springer Nature Switzerland AG
The registered company address is: Gewerbestrasse 11, 6330 Cham, Switzerland

Dedicated to OM SHAKTHI

Preface

Electrical signals are processed for controls, monitoring, transmission, and other requirements. The signals could be processed by analog or digital circuits. Digital signal processing (DSP) has distinct advantages over analog signal processing. Higher accuracy in controls, reduced noise in transmission and encryption are some of the advantages. DSP is highly popular in consumer, industrial, military and other applications. Understanding digital electronics is necessary to design logic circuits for digital signal processing.

This book presents the fundamentals of digital electronics for students to become DSP engineers. Most of the topics of digital electronics are presented in tutorial form for self-learning with high clarity. Digital signal processing (DSP) application information is provided for many topics of the subject to appreciate the practical significance of learning. Strong references are provided for additional learning. The presentation of the book is also structured for comfortable teaching for teachers.

Chapter 1 presents the overview of digital signal processing. Combinational logic circuits and related information are explained in Chaps. 2–6. Logic gates, logic minimization methods, digital hardware, hazards, binary number systems, binary codes, bit error detection, arithmetic operations, and arithmetic circuits are presented in the chapters.

Clock signal network and timing signals are used in computers and digital signal processing systems for controlling and performing their intended functions. They are explained in Chap. 7. Sequential logic elements (latches and flip-flops) and circuits (registers and counters) are presented in Chaps. 8–10.

Signal conversion architectures (ADC and DAC) are explained in Chap. 11. Simple programmable logic devices (SPLDs), complex programmable logic devices (CPLDs), and field programmable gate arrays (FPGAs) are presented in Chap. 12. The design of sequential logic circuits using Moore and Mealy machines is illustrated with examples in Chap. 13. The technologies and general application reliability information of digital ICs are presented in Chap. 14.

Bengaluru, India   Dhanasekharan Natarajan

Acknowledgements

Throughout my career, I was assisted by a team of committed and talented engineers and technicians with a lot of enthusiasm and initiatives. I would like to thank the team first for making me a better professional with time. I thank my wife, Rameswari, also for encouraging me to engage in writing books after my retirement.

Most importantly, I wish to express my sincere thanks to Prof. B. I. Khodanpur, HOD (CSE, Retired) of R V College of Engineering, one of the premier institutions in Bangalore. He provided me the opportunity to teach digital electronics to undergraduate students. I would also like to acknowledge the collaborative support from my colleagues at RVCE in teaching the subject.

Contents

About the Author

Dhanasekharan Natarajan, Electronics Engineer from College of Engineering, Guindy (presently Anna University), Chennai in 1970, obtained his post-graduate in Engineering Production (Q&R Option) from the University of Birmingham, UK, in 1984. He is Life Senior Member, IEEE (00064352), and his biography is published by Marquis, USA, in their fourth edition, *Who's Who in Science & Engineering*.

He retired as Assistant Professor in R V College of Engineering, Bengaluru. His earlier assignments were at Bharat Electronics and Radiall Protectron. His achievements in the industries include application of reliability techniques for defense equipment, root cause analysis on electronic component failures, qualification testing of electronic components as per USA and Indian military standards, designing and implementing computerized quality management system, designing software for optical interferometer, and design and manufacturing of lumped, semi-lumped, and microwave cavity filters using self-developed software.

He has authored three books, and the books are published by Springer. The first book, *A Practical Design of Lumped, Semi-lumped and Microwave Cavity Filters*, Springer, 2013, presents the design of L-C, tubular, and combline/iris-coupled cavity filters with examples.

The second book, *Reliable Design of Electronic Equipment: An Engineering Guide*, Springer, 2015, presents the application of derating, FMEA, overstress analysis, and reliability evaluation tests for designing reliable electronic equipment.

The third book, *ISO 9001 Quality Management Systems*, Springer, 2017, explains the requirements of ISO 9001 with examples from industries. Integrating QMS requirements with ERP software is the most effective method of implementing the system in organizations. The methods of integrating QMS requirements with ERP software are illustrated with examples. A chapter is dedicated for explaining the application of Indian classic, Thirukkural, for QMS planning.

Chapter 1
Overview of Digital Signal Processing

Abstract Electronic circuits are designed to process the functional requirements of product. Requirements such as controlling, monitoring and filtering could be processed using analog or digital signals. Analog and digital signal processing operations are presented for Switched-Mode Power Supplies (SMPS) using simplified functional block diagrams. Boolean algebra, De Morgan laws, Shannon theorems and their applications are also presented. Basic characteristics of digital signal are defined.

1.1 Types of Signals

Signals are broadly classified as analog and digital signals. Analog signals exist in nature. The sensory organs of humans recognize analog signals. Examples of analog signals are sound and temperature. Electrical supply voltages and currents are also examples of analog signals. Generally, the voltage level of analog signals varies with time.

Digital signals have only two voltage levels, namely, high and low. The signals exist in electronic circuits. In a transistor switching circuit, the collector-emitter voltage is low when the transistor is driven to saturation and the voltage is high when the transistor is non-conducting.

1.1.1 Analog Signal

Analog signal is represented graphically, relating its amplitude with time. Speech signal, captured by a microphone, is an example of analog signal and it is shown in Fig. 1.1. The analog signal is aperiodic. Its frequency and amplitude varies with time.

© Springer Nature Switzerland AG 2020
D. Natarajan, *Fundamentals of Digital Electronics*,
Lecture Notes in Electrical Engineering 623,
https://doi.org/10.1007/978-3-030-36196-9_1

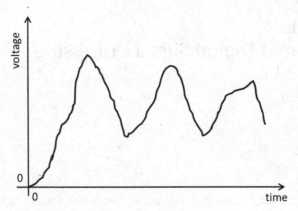

Fig. 1.1 Speech signal as a function of time

1.1.1.1 Sinusoidal Signal

Sinusoidal signal is another simple example of analog signal with positive and negative amplitude values varying continuously with time. The analog signal is periodic. The representation of the 1 V peak, 20 Hz sinusoidal signal is shown in Fig. 1.2. DC voltage is also analog signal. Its voltage level remains constant with time.

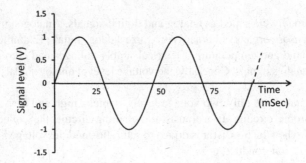

Fig. 1.2 Sinusoidal analog signal, 20 Hz

1.1.2 Digital Signal

Digital signal is used for control and data processing applications. The basic representation of digital signal is same for all applications. Unlike analog signal, digital signal has only two discrete voltage levels. Digital signal has one discrete voltage level at any instant of time. A 5 V digital signal has 0 V or 5 V at any instant of time. A 3.3 V digital signal has 0 V or 3.3 V at any instant of time. The lower voltage, 0 V is defined as logic Low or bit 0. The higher voltage is defined as logic High or bit 1.

Digital signal is also called binary signal as it has only two levels. The representation of digital signal is shown in Fig. 1.3.

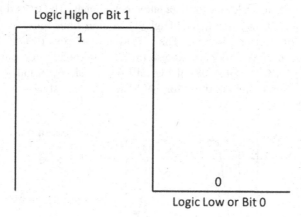

Fig. 1.3 Binary levels in digital signal

1.1.2.1 Data Signal

The digital signal, shown in Fig. 1.3, is 2-bit signal. Generally, data signal has four or more bits for digital processing applications. Data signal with more number of bits represent analog signal with higher resolution. The representation of 8-bit data signal, 11010101, is shown in Fig. 1.4. Generic names are also used for data signal. 4-bit data signal is called nibble. 8-bit data signal is called byte. Data signal with more than eight bits is called word.

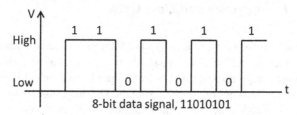

Fig. 1.4 Representation of 8-bit digital signal

1.2 Basic Characteristics of Digital Signal

Data signal and clock signal are the two categories of signals used for digital signal processing. Clock signals control data processing operations. The basic characteristics of digital signal are rise time, fall time, period, frequency and duty cycle. The characteristics of digital signals are defined.

1.2.1   Rise and Fall Times

Rise and fall times are relevant for data and clock signals. The transition of digital signal from logical Low (0) to logical High (1) requires finite time, considering the characteristics of electronic circuits. The transition time from 10% of logical Low to 90% of logical High is defined as rise time (t_r). Similarly, the transition time from 90% of logical High to 10% of logical Low is defined as fall time (t_f). The representations of rise and fall times for 5 V logic signal are shown in Fig. 1.5.

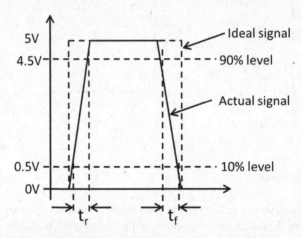

Fig. 1.5 Rise and fall times of 5 V digital signal

1.2.2   Period, Frequency and Duty Cycle

Period, frequency and duty cycle are relevant for clock signals. Symmetrical and asymmetrical clock signals are used in signal processing circuits and the signals are shown in Fig. 1.6a, b respectivly. Logical High (1) and logical Low (0) appear alternatively in both the types of the clock signals. The durations of logical High and logical Low are marked of t_H and t_L in the figure. They are equal in symmetrical signal. Period is the sum of the durations of logical High and logical Low. The expressions for period (T), frequency (F) and duty cycle are given below.

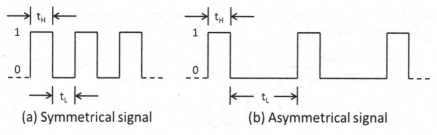

Fig. 1.6 Symmetrical and asymmetrical clock signals

$$Period, T = t_H + t_L$$
$$F, Frequency = \frac{1}{T}$$
$$Duty\ cycle(\%) = \frac{t_H}{T} \times 100$$

1.2.3 Signal Processing

Electronic circuits are designed to process the functional requirements of product. Requirements such as controlling, monitoring and filtering could be processed using analog or digital signals. Analog and digital signal processing operations are presented for Switched-Mode Power Supplies (SMPS) using simplified functional block diagrams.

1.3 Analog Signal Processing

Switched-mode power supplies (SMPS) are popular due to their power conversion efficiency. Overview of analog signal processing is presented for the control circuit of switched-mode power supplies (SMPS). Power MOSFET is in series with input supply and output load. The requirements of static and dynamic regulation of inputs or outputs under a range of operating conditions and with minimum loss of power are accomplished by actively controlling the on/off states of the power semiconductor devices [1].

The functional block diagram for regulating the output voltage of SMPS using analog controller is shown in Fig. 1.7. The input voltage for the control circuit is the error voltage obtained by comparing the output load voltage of MOSFET with a reference voltage. The output of the control circuit dynamically adjusts the gate voltage of power MOSFET to maintain regulated DC output voltage. The regulation control circuit uses op-amps, ICs, discrete semiconductors and passive components as required by circuit design.

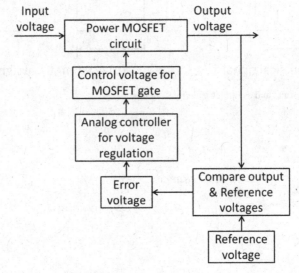

Fig. 1.7 Block diagram of SMPS with analog controller

1.4 Digital Signal Processing

Mathematical logic is the foundation of digital signal processing. A summary of the paper by Janet Heine Barnette on the application of mathematical logic in digital circuit design is presented [2]. On virtually the same day in 1847, two major new works were published by prominent mathematicians: *Formal Logic* by Augustus De Morgan (1806–1871) and *The Mathematical Analysis of Logic* by George Boole (1815–1864). In contrast to De Morgan, Boole took the significant step of explicitly adopting algebraic methods and developed a system of symbols $(\times, +)$ representing operations in 1854. Claude Shannon described the first engineering application of symbolic Boolean algebra in the paper, *A Symbolic Analysis of Relay and Switching Circuits*, in 1938 and the paper was based on his master's thesis at MIT, USA. Simplifications of expressions using Boolean algebra and synthesis of circuits were developed leading to new applications in digital signal processing.

1.4.1 Advantages

With the technological advancements in the design and manufacturing of ICs, digital signal processing (DSP) is widely used for many applications. Multi-voltage power supply and multi-utility measuring equipment with higher precision could be realized by user defined programs without changing hardware. Digital circuits are stable for wide range of temperature as the passive electronic components are replaced by digital hardware. However, DSP generates relatively higher level of electro-magnetic interference compared to analog signal processing. The advantages of DSP should be weighed against analog signal processing for deciding design options. Overview of digital signal processing is presented for digital switched-mode power supply (SMPS).

1.4.2 Digital SMPS

Digital SMPS is popular due to its programmability in most of the electronic applications. As in analog SMPS, power MOSFET is in series with input supply and output load. Digital signal processing is used for the control circuit of SMPS to achieve voltage regulation requirements. Digital hardware components are used for the control circuit. Digital controller may include fast, small analog-to-digital converters, hardware-accelerated programmable compensators, programmable digital modulators with very fine time resolution, and a standard microcontroller core to perform programming, monitoring and other system interface tasks [1]. The simplified functional block diagram for the digital SMPS using digital hardware is shown in Fig. 1.8.

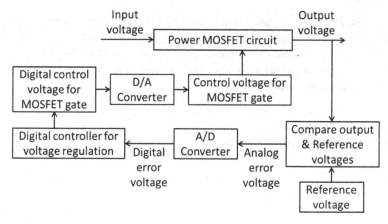

Fig. 1.8 Block diagram of SMPS with digital controller

1.4.3   Digital Hardware

In a broader sense, a digital system is implemented as a combination of digital hardware and software [3]. The requirements of digital circuit are expressed as logic function using the functional parameters of the circuit. The logic function of the circuit is implemented by using appropriate digital hardware components. Advanced ICs with many features are readily available for various DSP applications. However, it is necessary to understand the operation of various digital hardware components. The operation of the components are explained and illustrated with examples in the subsequent chapters. It is also necessary to simplify the logic functions of digital circuit before implementation.

1.5   Simplifying Logic Functions

Logic functions are simplified using Boolean algebra, De Morgan's laws and Shannon's expansion theorem. The laws and theorems present standardized logical relations. The standardized relations are used for:

(i)   Expressing the logic functions of digital circuits in standard form for simplification
(ii) Simplifying the logic functions before implementation to minimize digital hardware.

Variables, A, B, etc. are used in representing the standardized logical relations. The variables can have one of the two states, 1 or 0. Boolean algebra, De Morgan's laws and Shannon's theorems are presented after defining basic logical operators. The relevant laws and theorems are also used for explaining the operation of various digital hardware components.

1.5.1   Basic Logical Operators

There are three basic logical operators. They are logical sum, logical multiplication and logical NOT. Logical sum is represented by the plus operator. The logical value or the output of the expression $(A + B)$ is either 1 or 0. If the value of either A or B is equal to 1, the output of the expression is 1; otherwise the output is 0. The plus operator represents logical OR operation.

The dot operator represents logical multiplication. The logical value or the output of the expression $(A \cdot B)$ is either 1 or 0. If both the values of A and B are equal to 1, the output of the expression is 1; otherwise the output is 0. The dot operator represents logical AND operation.

Logical NOT operation is defined for one variable. It is represented as A'. It is also represented with a bar over the variable. If A is equal to 0, the output of the expression is 1; if A is equal to 1, the output is 0.

1.5.2 Boolean Algebra

The laws of Boolean algebra are listed below.

Distributive laws:

$$A \cdot (B + C) = (A \cdot B) + (A \cdot C)$$
$$A + (B \cdot C) = (A + B) \cdot (A + C)$$

Associative laws:

$$A + (B + C) = (A + B) + C$$
$$A \cdot (B \cdot C) = (A \cdot B) \cdot C$$

Commutative laws:

$$A + B = B + A$$
$$A \cdot B = B \cdot A$$

Inverse laws:

$$A + A' = 1$$
$$A \cdot A' = 0$$

Zero and one laws:

$$A + 1 = 1$$
$$A \cdot 0 = 0$$

Identity laws:

$$A + 0 = A$$
$$A \cdot 1 = A$$

Other relations for simplification:

$$A + (A \cdot B) = A$$
$$A \cdot (A + B) = A$$

$$A + A = A$$
$$A'' = A$$
$$A + A' \cdot B = A + B$$
$$A \cdot (A' + B) = A \cdot B$$
$$A \cdot B + A \cdot B' = A$$
$$(A + B) \cdot (A + B') = A$$
$$A \cdot B + A' \cdot C + B \cdot C = A \cdot B + A' \cdot C$$
$$(A + B) \cdot (A' + C) \cdot (B + C) = (A + B) \cdot (A' + C)$$

1.5.3 De Morgan Laws

De Morgan laws are shown for two variables. The laws are applicable for more than two variables also.

$$(A + B)' = A' \cdot B'$$
$$(A \cdot B)' = A' + B'$$

1.5.4 Shannon Theorems

Shannon theorems are shown for two variables. The theorems are applicable for more than two variables also.

$$F(A, B) = A' \cdot F(0, B) + A \cdot F(1, B)$$
$$F(A, B) = \left[A' + F(1, B) \right] \cdot [A + F(0, B)]$$

1.5.5 Simplification of Logic Functions

Three examples are provided for simplifying logic functions using Boolean algebra and De Morgan laws. The simplified logic functions could be verified by assigning 0 and 1 to the variables of the functions.

Example-1:

$$Z = AB'C + ABC + A'B'C' + AB'C$$

$$= AC(B' + B) + B'C'(C' + C)$$
$$Z = AC + B'C'$$

Example-2:

$$Z = A'(B + C') + A'B + A'C$$
$$= A'B + A'C' + A'B + A'C$$
$$= A'B + A'(C' + C)$$
$$= A'B + A'$$
$$= A'(B + 1)$$
$$Z = A'$$

Example-3:

Prove that $A(A' + B)' = AB'$

$$A(A' + B)' = A\left[(A')' \cdot B'\right]$$
$$= A[A \cdot B']$$
$$= A \cdot A \cdot B'$$
$$A(A' + B)' = AB'$$

1.6 Hardware Description Language

Logic circuits for various applications could be designed and developed in the traditional method or by using software simulation packages. In the traditional method, hardware components are assembled as per preliminary design and the circuit is tested for compliance to specifications. If the specifications are not met, design iterations are performed until compliance is achieved. The obvious drawbacks of the method are excessive cost and time to roll out new designs.

Software simulation packages offer more efficient method of designing digital logic circuits compared to the traditional method. The generic name for the simulation package is hardware description language (HDL). Verilog and VHDL (Very high speed integrated circuit HDL) are the two widely used hardware description languages.

Commercial sources are available for Verilog and VHDL software packages. Most of the commercial sources provide student edition of the software packages free for learning purposes. With the familiarity of C programming language, students could use the software packages for understanding the operation of digital hardware components and designing simple logic circuits. Help menus and user manuals of the software provide additional assistance for learning. Hardware description languages are not discussed in the book.

References

1. Maksimović D, Zane R, Erickson R (2004) Impact of digital control in power electronics. In: IEEE proceedings of the 16th international symposium on power semiconductor devices and ICs
2. Barnette JH (2013) Application of boolean algebra: claude shannon and circuit design. Colorado State University
3. Proakis J, Dimitris (1996) Digital signal processing: principles, algorithms, and applications. Prentice Hall

Chapter 2
Logic Gates

Abstract Logic gates are the basic building blocks of digital hardware. The operation of basic logic gates (OR, AND and NOT gates), universal logic gates (NOR and NAND gates), AND-OR-INVERTER, XOR and XNOR gates are explained. Active High and active Low input signals are defined and illustrated with examples.

2.1 Introduction

Logic gates are the basic building blocks of digital hardware. The following categories of logic gates are explained:

(i)   Basic logic gates
(ii) Universal logic gates
(iii) AND-OR-INVERT, XOR and XNOR gates.

2.2 Basic Logic Gates

Transistor switching circuits that perform logical OR, AND and NOT operations are grouped as basic logic gates. Standard ICs are available for the basic logic gates. The datasheets of the ICs could be referred for the transistor switching circuit diagrams of OR, AND and NOT gates. The operations of the basic logic gates are presented using Boolean algebra, truth table and timing diagram. The standardized symbols for the gates are also indicated.

2.2.1 OR Gate

OR gate has two or more input variables and the gate has one output. OR gate with two input variables (A & B), the output (Z) and the standardized symbol of the gate are shown in Fig. 2.1. The inputs and output are related by,

© Springer Nature Switzerland AG 2020
D. Natarajan, *Fundamentals of Digital Electronics*,
Lecture Notes in Electrical Engineering 623,
https://doi.org/10.1007/978-3-030-36196-9_2

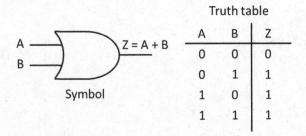

Fig. 2.1 OR gate with two inputs.

$$Z = A \, OR \, B = A + B$$

2.2.1.1   Truth Table

Truth table shows the relationship between the inputs and output of OR gate for all the combination of input variables. The output of OR gate is tabulated for the combinational inputs in truth table using Boolean algebra. The number of combinational inputs (c) is related to the number of input variables (n) to OR gate and it is given by:

$$c = 2^n$$

If the number of input variables (n) is two, the number of combinational inputs is four. If the number of input variables (n) is three, the number of combinational inputs is eight. The truth table for 2-input OR gate is shown in Fig. 2.1. The truth table for 3-input OR gate is shown in Fig. 2.2.

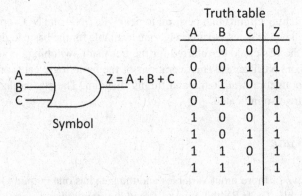

Fig. 2.2 OR gate with three inputs

2.2.1.2 Timing Diagram

Timing diagram shows the relationship between the inputs and output of OR gate graphically. The timing diagram for 2-input OR gate is shown in Fig. 2.3. The output is High when the input of A or B is High; otherwise it is Low.

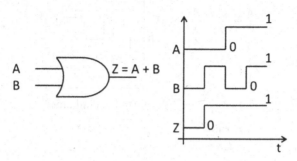

Fig. 2.3 Timing diagram for OR gate

2.2.1.3 Integrated Circuits

ICs for OR gates are available in TTL (Transistor Transistor Logic) and CMOS (Complementary Metal Oxide Semiconductor) technologies with many configurations. 74LS32 is an example of quad 2-input OR gate in TTL technology. 74HC4075 is an example of triple 3-input OR gate in CMOS technology. The data sheets of manufacturers provide logic circuit diagrams and application information.

2.2.2 AND Gate

AND gate has two or more input variables and the gate has one output. ICs (Ex.: 7408 and 74HC08) for AND gate are available with many configurations. The standardized symbol for AND gate with two inputs (A & B), output (Z), truth table and timing diagram are shown in Fig. 2.4. The output is High when both the inputs of A and B are High; otherwise it is Low. The inputs and output are related by,

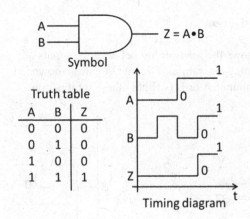

Symbol

Truth table

A	B	Z
0	0	0
0	1	0
1	0	0
1	1	1

Timing diagram

Fig. 2.4 AND gate with two inputs

$$Z = A \, AND \, B = A \cdot B$$

Truth table for AND gate with more than two input variables could also be prepared. The inputs and output of 3-input AND gate are related by,

$$Z = A \cdot B \cdot C$$

2.2.3 NOT Gate (Inverter)

NOT gate is generally known as inverter. It has one input and one output. The standardized symbol for inverter, the input (A), the output (Z), truth table and timing diagram of the gate are shown in Fig. 2.5. The output is High when the input is Low. The output is Low when the input is High. ICs (Ex.: 7404 and CD4069) for inverter are available with many configurations. The input and output are related by,

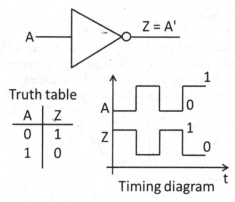

Fig. 2.5 Inverter with input and output signals

$$Z = A'$$

2.2.4 Active High and Active Low Input Signals

A logic gate is said to be asserted when logic 1 signal is applied to the input port of the gate. The logic 1 signal is called active High input signal to the logic gate. The application of active High input signals to 2-input AND gate is shown in Fig. 2.6a. The output of the AND gate is 1 when A and B are equal to 1. Active High input signal is also called positive logic input signal.

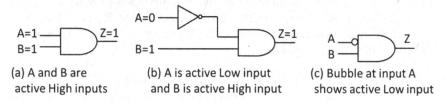

(a) A and B are active High inputs　　(b) A is active Low input and B is active High input　　(c) Bubble at input A shows active Low input

Fig. 2.6 Active High and Active Low assertions

　　The AND gate in Fig. 2.6a could be asserted by applying logic 0 signal through inverter to the input of the gate. The schematic diagram for asserting the 2-input AND gate with logic 0 signal at A and logic 1 (active High) signal at B is shown in Fig. 2.6b. The output of the AND gate is 1 with A = 0 and B = 1. The input logic 0 signal at A is called active Low input signal to the AND gate. Active Low input signal is also called negative logic input signal.

　　The general representation of 2-input AND gate with active High and Low inputs is shown in Fig. 2.6c. The active Low input signal is indicated with a bubble at the input of the AND gate. The inverter is omitted in the representation.

2.2.4.1 Bubbled Output

2-input AND gate is shown with bubble at the output of gate in Fig. 2.7. It indicates the output of the gate is inverted i.e. active Low signal. The datasheets of ICs for digital hardware use bubbles or triangles for indicating active Low signal at the inputs or outputs, as applicable. The triangle symbol is as per IEEE and IEC standards.

Fig. 2.7 AND gate with bubble at its output

2.3 Universal Logic Gates

Universal logic gates are NOR and NAND gates. Basic logic gates OR, AND and NOT gates could be realized from the universal gates. The operation of universal logic gates and the realization of the basic logic gates using the universal logic gates are presented.

2.3.1 NOR Gate

The standard NOR gate is OR gate followed by inverter i.e. OR gate with bubble at its output. Two or more inputs are associated with NOR gate and the gate has one output. The standardized symbol for NOR gate with two inputs (A & B), the output (Z) and the truth table are shown in Fig. 2.8. The output of NOR gate is 1 when A $= B = 0$; otherwise, it is 0. ICs for standard NOR gate are available. The inputs and output of NOR gate are related by,

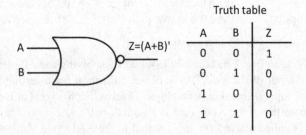

Fig. 2.8 Standard NOR gate with two inputs

$$Z = (A + B)'$$

2.3.1.1 Bubbled AND Gate

NOR gate could also be realized by having inverters for the inputs of AND gate i.e. bubbled AND gate. The standardized symbol for bubbled AND gate with two inputs (A & B), the output (Z) and the truth table of the gate are shown in Fig. 2.9. The inputs and output of bubbled NOR gate are related by,

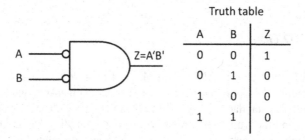

Truth table		
A	B	Z
0	0	1
0	1	0
1	0	0
1	1	0

Fig. 2.9 Bubbled AND gate with two inputs

$$Z = A' \cdot B'$$

The outputs of standard NOR gate and bubbled AND gate are same for the same combination of inputs. It could be proved by applying De Morgan's theorem on the outputs.

$$Z = A' \cdot B' = (A + B)'$$

2.3.1.2 Basic Logic Gates from NOR Gate

The three basic logic gates, NOT, OR and AND, could be realized from NOR gate. The logic circuits for realizing the gates are shown in Fig. 2.10. The derivations for the output expressions for the gates are also shown in the figures.

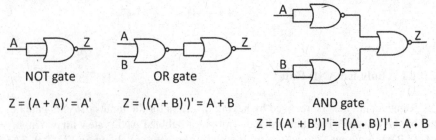

Fig. 2.10 Basic logic gates from NOR gate

2.3.2 NAND Gate

The standard NAND gate is AND gate followed by inverter i.e. AND gate with bubble at its output. Two or more inputs are associated with NAND gate and the gate has one output. The standardized symbol for NAND gate with two inputs (A & B), the output (Z) and the truth table are shown in Fig. 2.11. ICs for standard NAND gate are available. The inputs and output of NAND gate are related by,

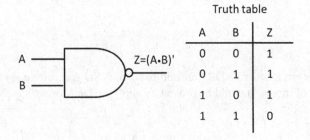

Fig. 2.11 Standard NAND gate with two inputs

$$Z = (A \cdot B)'$$

2.3.2.1 Bubbled OR Gate

NAND gate could also be realized by having inverters for the inputs of OR gate i.e. bubbled OR gate. The standardized symbol for bubbled OR gate with two inputs (A & B), the output (Z) and the truth table are shown in Fig. 2.12. The inputs and output of bubbled OR gate are related by,

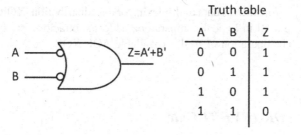

Truth table

A	B	Z
0	0	1
0	1	1
1	0	1
1	1	0

Fig. 2.12 Bubbled OR gate with two inputs

$$Z = A' + B'$$

The outputs of standard NAND gate and bubbled OR gate are same for the same combination of inputs. It could be proved by applying De Morgan's theorem on the outputs.

$$Z = A' + B' = (A \cdot B)'$$

2.3.2.2   Basic Logic Gates from NAND Gate

The three basic logic gates, NOT, OR and AND, could be realized from NAND gate. The logic circuits for realizing the gates are shown in Fig. 2.13. The derivations for the output expressions for the gates are also shown in the figures.

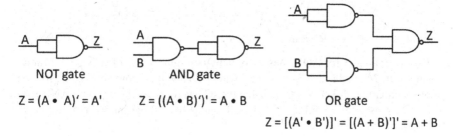

NOT gate   AND gate

$Z = (A \cdot A)' = A'$ $Z = ((A \cdot B)')' = A \cdot B$

OR gate

$Z = [(A' \cdot B')]' = [(A + B)']' = A + B$

Fig. 2.13 Basic logic gates from NAND gate

2.4   General Purpose Logic Gates

AND-OR-INVERT (AOI), XOR and XNOR are general purpose logic gates. They are used in many applications. Datasheets of AOI gate indicate that the device is used for data control, such as data multiplexing or data distribution. XOR and XNOR gate circuits are used in logical comparators, adders/subtractors, parity generators and checkers. The operation of AOI, expandable AOI, XOR and XNOR gates, are presented.

2.4.1 AND-OR-INVERT Gate

As the name implies, AND-OR-INVERT (AOI) gate circuit is formed by AND, OR and inverter gates. The schematic diagram of AOI is shown in Fig. 2.14a. AOI gate circuit contains a minimum of two AND gates and one NOR (OR followed by inverter) gate. AOI with two AND gates is designated as 2-wide, AOI with three AND gates is 3-wide and so on. Each AND gate could have two or more inputs. 2-input, 2-wide AOI is shown in Fig. 2.14a with Boolean equation.

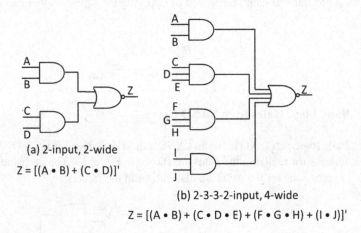

(a) 2-input, 2-wide

$$Z = [(A \bullet B) + (C \bullet D)]'$$

(b) 2-3-3-2-input, 4-wide

$$Z = [(A \bullet B) + (C \bullet D \bullet E) + (F \bullet G \bullet H) + (I \bullet J)]'$$

Fig. 2.14 AND-OR-INVERT gates

ICs are available for AOI gates with many configurations. 7451 is dual 2-input, 2-wide AOI. 74LS54 is 2-3-3-2-input, 4-wide AOI and it is shown in Fig. 2.14b with Boolean equation. 2-3-3-2 indicates that the IC has four AND gates; first AND gate has two inputs; the second and third AND gates have three inputs and the fourth AND gate has two inputs.

2.4.2 Expandable AND-OR-INVERT Gate

The number of AND gates in AOI gate is limited considering the complexity in the fabrication of ICs. Circuit applications might require higher number of AND gates than the available number of AND gates in ICs. Expandable AOI gate provides solution for the applications. The application of expandable AOI gate is illustrated with an example.

2.4.2.1 Components and Their Interconnection

Assume that circuit application requires 2-2-4-4-input, 4-wide AOI gate. Two types of ICs are required to satisfy the circuit needs. Expandable dual 2-input, 2-wide IC (7450) and dual 4-input expanders (7460) are used to realize the circuit needs.

Unlike general purpose AOI gate, the OR-Invert gate of expandable dual 2-input, 2-wide AOI gate (IC 7450) has two additional inputs, namely, bubble and arrow. The schematic of the expandable AOI gate is shown in Fig. 2.15a. The dual 4-input expander (IC 7460) has two 4-input AND gates. The AND gates of the expander have two outputs, namely, bubble and arrow. The schematic of the expander is shown in Fig. 2.15b. The bubble and arrow outputs of the AND gates of the expander are connected to the corresponding inputs of OR-Inverter of expandable AOI gate. The realization of 2-2-4-4-input, 4-wide, AOI gate is shown in Fig. 2.15c.

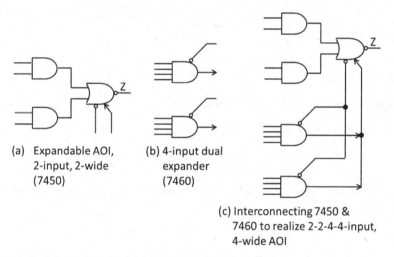

(a) Expandable AOI,
     2-input, 2-wide
     (7450)

(b) 4-input dual
    expander
    (7460)

(c) Interconnecting 7450 &
    7460 to realize 2-2-4-4-input,
    4-wide AOI

Fig. 2.15 Expandable AOI, 2-2-4-4-input, 4-wide, with dual expander

2.4.3 XOR Gate

Generally, standard ICs are available for XOR gates with two inputs. IC 7486 is a Quad 2-input XOR. The output of 2-input XOR gate is High when one of its two inputs (variables) is High; otherwise, the output of the gate is Low.

XOR gate could be implemented by using appropriate combination of inverter, AND, OR, NOR and NAND gates also. 2-input XOR gate could be conveniently implemented using Quad 2-input NAND gate (IC 7400). The NAND gate implementation of 2-input XOR gate, the symbol and the truth table of the gate are shown in Fig. 2.16. The operator, $\oplus$, represents XOR operation. The inputs and output of 2-input XOR function are related by,

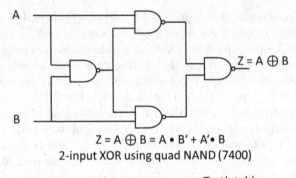

$$Z = A \oplus B = A \cdot B' + A' \cdot B$$
2-input XOR using quad NAND (7400)

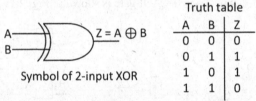

Symbol of 2-input XOR

Truth table

A	B	Z
0	0	0
0	1	1
1	0	1
1	1	0

Fig. 2.16 Implementation of 2-input XOR gate

$$Z = (A \oplus B) = A \cdot B' + A' \cdot B$$

The XOR identities are listed below. They could be verified using truth table or using Boolean equations.

$$A \oplus 0 = A$$
$$A \oplus 1 = A'$$
$$A \oplus A = 0$$
$$A \oplus A' = 1$$
$$A \oplus B' = A' \oplus B = (A \oplus B)'$$

2.4.3.1 XOR with Higher Order Inputs

XOR function for 3 or more input variables is implemented by cascading 2-input XOR gates in a tree format. The implementation of 3-input and 4-input XOR functions is shown in Fig. 2.17.

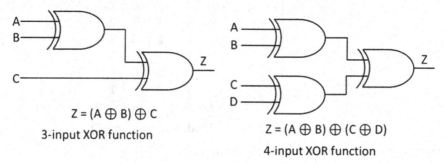

$$Z = (A \oplus B) \oplus C$$
3-input XOR function

$$Z = (A \oplus B) \oplus (C \oplus D)$$
4-input XOR function

Fig. 2.17 Implementation of 3-input & 4-input XOR functions

2.4.3.2 Modulo-2 Sum Bit

XOR gate generates output which is the modulo-2 sum of two binary inputs (digits). Modulo-2 sum is the sum of two digits, ignoring the resultant carry bit. The modulo-2 sum of $(1 + 1)$ is 0, ignoring the resultant carry bit, 1. The modulo-2 sum is indicated as output in truth table. Modulo-2 sum is explained for XOR function with three inputs.

The logic inputs of the variables of 3-input XOR function are shown in Fig. 2.18. Modulo-2 sum is applied for the logic inputs of the variables in stages. Modulo-2 sum is first applied for the variables, A and B, obtaining the intermediate output, X. The sum is applied again for the variables X and C to obtain the final output, Z. The final truth table obtained using the modulo-2 sum of the input variables is shown in Fig. 2.18.

Logic inputs of variables				Truth table using modulo-2 sum				
A	B	C	Z	A	B	X	C	Z
0	0	0		0	0	0	0	0
0	0	1		0	0	0	1	1
0	1	0		0	1	1	0	1
0	1	1		0	1	1	1	0
1	0	0		1	0	1	0	1
1	0	1		1	0	1	1	0
1	1	0		1	1	0	0	0
1	1	1		1	1	0	1	1

X = Modulo-2 sum, (A + B)

Z = Modulo-2 sum, (X + C)

Fig. 2.18 Truth table for 3-input XOR using modulo-2 sum

2.4.3.3   Odd Function

XOR functions are odd functions. The output of XOR function is 1 when the number of input variables having logic 1 is odd. The truth table of 3-input XOR function (Fig. 2.18) illustrates that XOR is odd function. The outputs of the XOR function are 1 for the inputs, 001, 010, 100 and 111. Similarly, the outputs of 2-input XOR gate are 1 for the inputs, 01, and 10.

2.4.4 XNOR Gate

XNOR (eXclusive NOR) gate is XOR gate followed by inverter. The output of XNOR gate is the complement of the output of XOR gate. Standard ICs (Ex. 74LS266, Quad 2-input XNOR) are available for XNOR gates with two inputs. The output of 2-input XNOR gate is High when both the inputs are either Low or High; otherwise, the output is Low.

XNOR gate could be implemented by using appropriate combination of inverter, AND, OR, NOR and NAND gates. 2-input XNOR gate could be conveniently implemented using Quad 2-input NOR gate (IC 7425). The NOR gate implementation, the symbol and the truth table of XNOR gate are shown in Fig. 2.19. The operator, $\odot$, represents XNOR operation. The inputs and output of 2-input XNOR function are related by,

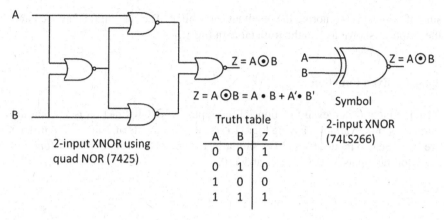

$Z = A \odot B = A \cdot B + A' \cdot B'$

Truth table

A	B	Z
0	0	1
0	1	0
1	0	0
1	1	1

2-input XNOR using quad NOR (7425)

Symbol

2-input XNOR (74LS266)

Fig. 2.19 Implementation of 2-input XNOR gate

$$Z = (A \odot B) = (A \oplus B)' = A \cdot B + A' \cdot B'$$

2.4.4.1 XNOR with Higher Order Inputs

XNOR function for 3 or more inputs (variables) is implemented by cascading 2-input XOR gates in a tree format except that the output XOR gate is replaced by XNOR gate. The implementation of 3-input and 4-input XNOR functions is shown in Fig. 2.20.

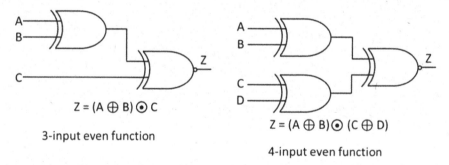

$Z = (A \oplus B) \odot C$

3-input even function

$Z = (A \oplus B) \odot (C \oplus D)$

4-input even function

Fig. 2.20 Implementation of 3-input & 4-input even functions

2.4.4.2 Inverted Modulo-2 Sum Bit

Like XOR gate, XNOR also generates output which is the modulo-2 sum of two binary inputs (digits) except that the modulo-2 sum bit is inverted. The modulo-2 sum of $(1 + 1)$ is 0 ignoring the resultant carry bit, 1. The sum bit 0 is inverted and the output is shown as 1 in the truth table in Fig. 2.19.

2.4.4.3   Even Function

XNOR functions are even functions. The output of XNOR function is 1 when the number of input variables having logic 1 is even. The outputs of 2-input XNOR gate are 1 for the two inputs, 00 and 11. Similarly, the outputs of 3-input XNOR function are 1 for the inputs, 000, 011, 101 and 110.

Chapter 3
Combinational Logic Minimization

Abstract The general approach for designing combinational logic circuits is illustrated with an example. Logic functions are the output of the design of combinational circuits and they should be simplified before implementation. Algebraic method, Karnaugh mapping and Quine-McCluskey methods for simplifying logic functions are explained and illustrated with examples. Hazards usually exist with the implementation of logic functions. The types of hazards and compensating circuits to eliminate the hazards are explained.

3.1 Overview of Combinational Logic Design

Combinational and sequential logic circuits are used in digital system. Combinational logic circuit is developed using basic, universal and other gates for various applications. Examples of combinational logic circuits are multiplexers, decoders, adders etc. Combinational circuits do not contain memory to store output data. The output of the circuits depends only on current inputs.

Sequential logic circuit contains combinational logic and memory elements. Flip-flop is an example of memory element. The memory element stores output data. The output of sequential circuits depends on current inputs as well as the stored outputs of the memory elements before the application of the inputs. Memory elements and sequential logic circuits are presented in subsequent chapters.

Logic functions are required to design combinational logic circuits. They are obtained from the functional requirements of electronic units. Logic functions are simplified before implementation to minimize digital hardware. Logic minimization techniques are illustrated with examples in this chapter. General design approach is explained for obtaining logic functions from the functional requirements of electronic units before presenting the minimization techniques.

© Springer Nature Switzerland AG 2020
D. Natarajan, *Fundamentals of Digital Electronics*,
Lecture Notes in Electrical Engineering 623,
https://doi.org/10.1007/978-3-030-36196-9_3

3.1.1 General Design Approach

The functional requirements of electronic units are the basic inputs for the design of combinational logic circuits. The basic design inputs are converted and represented in the form of truth table. Logic function is obtained from the truth table. The logic function is minimized i.e. simplified. The simplified logic function is implemented using appropriate gate circuits. The general approach for designing combinational logic circuits is illustrated with an example.

3.1.1.1 Functional Requirements

Assume that a digital control circuit needs to be designed for a motor circuit. Functional requirements of the motor circuit are usually in the form of simple statements. The motor circuit has dual controls (S_1 & S_2) for switching on the motor circuit. No damage to motor circuit occurs even when both S_1 and S_2 are switched on. The motor circuit has thermal protection. Motor circuit should function when S_1 or S_2 or both the switches are switched on and when the surface temperature (T) of the motor is less than the pre-set temperature level. The functional requirements of motor circuit are converted and represented in the form of truth table.

3.1.1.2 Preparing Truth Table

There are three variables (inputs) associated with the motor circuit and the variables are S_1, S_2 and T. It is assumed that inputs are converted into logic signals and the logic signals are readily available. The logic signal for S_1 or S_2 is High when they are switched on. The logic signal for T is High when surface temperature is within the pre-set level. Let Z represent the output of the digital control circuit. When the output (Z) of the control circuit is High, the motor circuit is switched on. The output (Z) is High when:

(i) S_1 and T are High.
(ii) S_2 and T are High.
(iii) S_1, S_2 and T are High.

The combinations of the inputs (S_1, S_2 and T) and the output (Z) of digital control circuit could be represented in the form of truth table. The truth table of the digital control circuit is shown in Fig. 3.1.

S_1	S_2	T	Z
0	0	0	0
0	0	1	0
0	1	0	0
0	1	1	1
1	0	0	0
1	0	1	1
1	1	0	0
1	1	1	1

Fig. 3.1 Truth table for motor circuit

3.1.1.3 Obtaining Logic Function

Logic function is obtained from the truth table of the function. Two standardized forms, namely, Sum-of-Products (SOP) and Product-of-Sums (POS) are available for expressing the logic function. The terms associated with SOP and POS forms and the methods of obtaining the logic function from truth table in both the forms are illustrated for the motor circuit in subsequent sections.

3.1.1.4 Simplifying Logic Function

Logic functions, obtained from truth table, generally require higher number of gates for implementation. They should be simplified to minimize the number of gate circuits for ensuring better performance and cost benefits. Propagation delay between input and output signals is low with minimum number of gates, resulting in faster performance. Three methods of simplifying logic function are presented in subsequent sections. The methods are illustrated with many examples for understanding the rules of simplifications. Simplifying the logic function of the motor circuit is also presented. The methods for simplifying logic function are:

- Algebraic method
- Karnaugh Mapping method
- Quine-McCluskey method.

3.1.1.5 Hazards

Simplified logic function is implemented using gate circuits. Hazards usually exist with the implementation of logic functions. Unwanted changes in the logic states of output are called hazards. The types of hazards are explained with timing diagrams and compensating circuits are presented to eliminate hazards.

3.2   Logic Function in SOP Form

Logic function is expressed as a logical sum of product terms in SOP′ form. For example, consider a 2-variable logic function,

$$Z = (A'B) + (AB')$$

$(A'B)$ and (AB') are the product terms in the expression. The output (Z) of the logic function is expressed as a sum of the two product terms in SOP form. It is necessary to define minterm before illustrating SOP method for obtaining the logic function of motor circuit.

3.2.1   Minterm

Minterm is defined as the logical product term containing all the variables of a function in complemented and uncomplemented form applicable to the logic status of the variables. The variables having logic level, 1 are not complemented and the variables having logic level, 0 are complemented in minterms. The minterms of logic function are identified as m_0, m_1, m_2, etc. The number of minterms of logic function is equal to 2^n, where n is number of the variables of the function.

For example, consider a logic function with two ($n = 2$) variables, A and B. There are four (2^2) minterms for the function. The minterms are:

For $A = 0$ and $B = 0$, $m_0 = A'B'$
For $A = 0$ and $B = 1$, $m_1 = A'B$
For $A = 1$ and $B = 0$, $m_2 = AB'$
For $A = 1$ and $B = 1$, $m_3 = AB$.

3.2.2   Obtaining Logic Function

The truth table of 3-variable motor circuit function is reproduced in Fig. 3.2. The number of minterms of the function is 2^3 i.e. eight. The minterms of the input variables are also shown in the truth table and they are serially numbered from m_0 to m_7. Logic function is the sum of the minterms having logic output (Z), 1. Referring to Fig. 3.2, Z is 1 for m_3, m_5 and m_7. The logic function of the motor circuit is given by the expression,

S_1	S_2	T	Z	Minterms
0	0	0	0	$m_0: S'_1 S'_2 T'$
0	0	1	0	$m_1: S'_1 S'_2 T$
0	1	0	0	$m_2: S'_1 S_2 T'$
0	1	1	1	$m_3: S'_1 S_2 T$
1	0	0	0	$m_4: S_1 S'_2 T'$
1	0	1	1	$m_5: S_1 S'_2 T$
1	1	0	0	$m_6: S_1 S_2 T'$
1	1	1	1	$m_7: S_1 S_2 T$

Fig. 3.2 Minterms of the motor circuit logic function

$$Z = m_3 + m_5 + m_7 = (S'_1 S_2 T) + (S_1 S'_2 T) + (S_1 S_2 T)$$

Alternatively, the expression is written with sigma (Σ) notation representing the sum of minterms, as:

$$Z = F(S_1, S_2, T) = \Sigma\, m(3, 5, 7)$$

3.3 Logic Function in POS Form

Logic function is expressed as logical product of sum terms in POS form. For example, consider a 2-variable logic function,

$$Z = (A' + B)(A + B')$$

$(A' + B)$ and $(A + B')$ are sum terms in the expression. The output (Z) of the logic function is expressed as a product of the two sum terms in POS form. Maxterm is defined before illustrating POS method for obtaining the logic function of motor circuit.

3.3.1 Maxterm

Maxterm is defined as the logical sum term containing all the variables of a function in complemented and uncomplemented form applicable to the logic status of the variables. The variables having logic level, 1 are complemented and the variables having logic level, 0 are not complemented in maxterms. The maxterms of logic function are identified as M_0, M_1, M_2, etc. The number of maxterms of logic function is equal to 2^n, where n is number of the variables of the function.

For example, consider a logic function with two ($n = 2$) variables, A and B. There are four (2^n) maxterms for the function. The maxterms are:

For $A = 0$ and $B = 0$, $M_0 = A + B$
For $A = 0$ and $B = 1$, $M_1 = A + B'$
For $A = 1$ and $B = 0$, $M_2 = A' + B$
For $A = 1$ and $B = 1$, $M_3 = A' + B'$.

3.3.2 Obtaining Logic Function

The truth table of 3-variable motor circuit function is reproduced in Fig. 3.3. The number of maxterms of the function is 2^3 i.e. eight. The maxterms of the input variables are also shown in the truth table and they are serially numbered from M_0 to M_7. Logic function is the product of the maxterms having logic output level (Z), 0. Referring to Fig. 3.3, Z is 0 for M_0, M_1, M_2, M_4 and M_6. The logic function of the motor circuit is given by the expression,

S_1	S_2	T	Z	Maxterms
0	0	0	0	M_0: $S_1 + S_2 + T$
0	0	1	0	M_1: $S_1 + S_2 + T'$
0	1	0	0	M_2: $S_1 + S'_2 + T$
0	1	1	1	M_3: $S_1 + S'_2 + T'$
1	0	0	0	M_4: $S'_1 + S_2 + T$
1	0	1	1	M_5: $S'_1 + S_2 + T'$
1	1	0	0	M_6: $S'_1 + S'_2 + T$
1	1	1	1	M_7: $S'_1 + S'_2 + T'$

Fig. 3.3 Maxterms of the motor circuit logic function

$$Z = M_0 M_1 M_2 M_4 M_6$$
$$Z = (S_1 + S_2 + T)(S_1 + S_2 + T')(S_1 + S'_2 + T)(S'_1 + S_2 + T)(S'_1 + S'_2 + T)$$

Alternatively, the expression is written with pi (Π) notation representing the product of maxterms, as:

$$Z = F(S_1, S_2, T) = \Pi M(0, 1, 2, 4, 6)$$

3.4 Algebraic Method for Logic Simplification

Standard Boolean expressions are used for simplifying logic functions in the algebraic method. The algebraic method is suitable for functions with small number of variables. Simplification of the logic functions in SOP form and in POS form for the motor circuit example is illustrated.

3.4.1 Simplifying Logic Function in SOP Form

The logic function of motor circuit, obtained from SOP method in Sec. 3.2, is shown below. The implementation of the function before simplification is shown in Fig. 3.4 for information. It requires two inverters, three AND gates and one OR gate.

$$Z = (S_1'S_2T) + (S_1S_2'T) + (S_1S_2T)$$

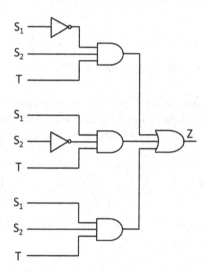

Fig. 3.4 Implementation of logic function before simplification

3.4.1.1 Simplification

Appropriate Boolean expressions are used for simplifying the function,

$$Z = (S_1'S_2T) + (S_1S_2'T) + (S_1S_2T)$$
$$Z = T(S_1'S_2) + T(S_1S_2') + T(S_1S_2)$$
$$Z = T[(S_1'S_2) + (S_1S_2') + (S_1S_2)]$$

$$Z = T\big[(S_1'S_2) + (S_1S_2') + (S_1S_2) + (S_1S_2)\big]$$
$$Z = T\big[(S_1S_2') + (S_1S_2) + (S_1'S_2) + (S_1S_2)\big]$$
$$Z = T\big[(S_1S_2) + (S_1S_2') + (S_1S_2) + (S_1'S_2)\big]$$
$$(S_1S_2) + (S_1S_2') = S_1 \text{ and } (S_1S_2) + (S_1'S_2) = S_2$$
$$Z = T(S_1 + S_2)\text{n}$$

The implementation of the simplified logic function is shown in Fig. 3.5. The implementation of the logic function requires just one each of AND and OR gates. The efficacy of the gate circuit could be verified using the truth table of the motor circuit.

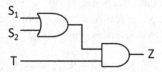

Fig. 3.5 Implementation of logic function after simplification

3.4.2 Simplifying Logic Function in POS Form

The logic function of motor circuit, obtained from POS method in Sec. 3.3, is reproduced and appropriate Boolean expressions are used for simplifying the function. It could be observed that the simplified function is same as that of SOP form in Sec. 3.4.1.1.

$$Z = (S_1 + S_2 + T)(S_1 + S_2 + T')(S_1 + S_2' \cdot T)(S_1' + S_2 + T)(S_1' + S_2' + T)$$
$$(S_1 + S_2 + T)(S_1 + S_2' \cdot T) = (S_1 + T + S_2)(S_1 + T + S_2') = (S_1 + T)$$
$$(S_1' + S_2 + T)(S_1' + S_2' + T) = (S_1' + T + +S_2)(S_1' + T + S_2') = (S_1' + T)$$
$$Z = (S_1 + T)(S_1 + S_2 + T')(S_1' + T)$$
$$(S_1 + T)(S_1' + T) = T$$
$$Z = T(S_1 + S_2 + T')$$
$$Z = T(T' + S_1 + S_2)$$
$$Z = TT' + T(S_1 + S_2) = T(S_1 + S_2)$$

3.4.3 Transformation Between SOP and POS Forms

Logic functions in SOP and POS forms are equivalent. Each form could be transformed into the other form. For example, consider the 3-variable logic functions of motor circuit. The SOP and POS forms are equated as,

$$Z = F(S_1, S_2, T) = \Sigma\, m(3, 5, 7) = \Pi\, M(0, 1, 2, 4, 6)$$

In general, the missing state numbers of one form are indicated in the other form. The SOP form contains the states, 3, 5 and 7 and the POS form contains the states, 0, 1, 2, 4 and 6. Logic function could be obtained in the form that gives minimum number of terms for simplification.

3.5 Karnaugh Mapping

Karnaugh mapping method is named after the American physicist, Maurice Karnaugh, who developed the method in 1954 at Bell Labs. Karnaugh mapping (abbreviated as K-map) is a graphical method of obtaining simplified logic functions directly from truth table without using Boolean expressions. The application of K-map is illustrated with many examples for obtaining simplified logic function in SOP form. One example is provided to obtain logic function in POS form. Variable-entered K-map and Don't care conditions in K-map are also illustrated with examples. Generally, the application of K-map is limited to logic functions with a maximum of five variables.

3.5.1 K-map with Two Variables

K-map has cells to place the outputs of truth table. The number of cells is equal to the number of minterms of the truth table. Truth table for two variables is shown in Fig. 3.6a. The K-map of the truth table has four cells for the four minterms of the truth table. The cells are identified by labelling the rows and columns of the K-map.

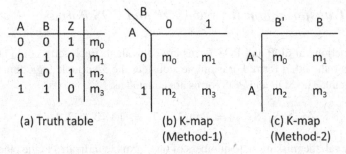

Fig. 3.6 General representation of K-map for 2-input function

3.5.1.1 Labelling and Filling Cells

The variables of logic function are used for labeling the rows and columns of K-map. Labelling the cells of K-map could be done in two ways and the methods are shown in Fig. 3.6b, c. The locations of placing the outputs of the truth table in the cells of the K-map are indicated by the minterms of the truth table in the figures. As the variables having logic level of 0 are complemented in minterms, the method shown in Fig. 3.6c is used in the book. The method is also relatively more convenient to write the outputs of K-map i.e. simplified logical function using the method.

The truth table is shown in Fig. 3.7a. The 1 s of the truth table are first placed at the appropriate locations of K-map. The output of the truth table is 1 for m_0 and m_2. Accordingly, 1 is placed in the cells, $A'B'$ and AB' of the K-map. The remaining cells of the K-map are filled with 0. The K-map is shown in Fig. 3.7b.

A	B	Z
0	0	1
0	1	0
1	0	1
1	1	0

(a) Truth table

	B'	B
A'	1	0
A	1	0

(b) K-map

Fig. 3.7 K-map for 2-variable function

3.5.1.2 Looping of Adjacent 1-Cells

Adjacent 1-cells (cells with 1) are looped as pairs, quads and octets. The K-map in Fig. 3.7b has one pair of adjacent 1 s and the looped pair is shown in Fig. 3.8b. The truth table is also reproduced for reference in the figure.

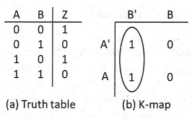

(a) Truth table (b) K-map

Fig. 3.8 Looping of adjacent 1 s in K-map

3.5.1.3 Obtaining Simplified Logic Function from K-Map

The simplified logic function is obtained from the two minterms of the loop in Fig. 3.8b. The minterms are $A'B'$ and AB'. The variable that differs by one bit in the minterms is discarded. A' and A differ by one bit in the minterms and they are discarded. The simplified logic function in SOP form is:

$$Z = B'$$

3.5.1.4 Example 2: K-map with Two Variables

The truth table of a function with two variables and its representation in K-map are shown in Fig. 3.9a, b respectively. There are two pairs adjacent 1-cells. The vertical and horizontal loops are shown in the K-map in the figure. The two loops overlap each other and overlapping of loops is permitted. Diagonal looping of adjacent 1-cells is not permitted. The variables, B' and B, differ by one bit in the horizontal loop and they are discarded. The output of the horizontal loop is A'. The variables, A' and A, differ by one bit in the vertical loop and they are discarded. The output of the vertical loop is B'. The simplified logic function in SOP form is:

$$Z = A' + B'$$

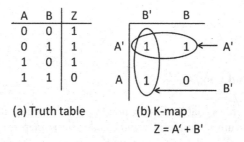

(a) Truth table (b) K-map

$$Z = A' + B'$$

Fig. 3.9 Example 2: K-map for 2-variable function

3.5.2 K-map with Three Variables

The truth table of functional requirement with three variables has eight output states and eight minterms, m_0 to m_7. K-map for three variables requires eight cells for placing the eight outputs of the truth table. The K-map could be drawn with two rows and four columns or with four rows and two columns. Gray code is used for labeling the rows and columns with the variables of truth table.

3.5.2.1 Gray Code for Labelling the Rows and Columns of K-map

Truth table uses BCD (binary coded decimal) system and the successive codes represent decimal numbers in ascending order. For example, the truth table with three variables represents decimal numbers from 0 to 7 in ascending order. In Gray code, the successive codes change by one bit.

Gray code was invented in 1947 by Frank Gray, a researcher at Bell Labs, and a patent was awarded for the invention in 1953 [1]. Two bit Gray code is shown in Fig. 3.10a and the same order is used for labelling K-map with three variables. If K-map with four rows and two columns are used, the variables, A and B, are used to label the rows. If K-map with two rows and four columns are used, the variables, B and C, are used to label the columns. The order of labelling the variables to rows and columns are shown in Fig. 3.10b, c. The minterms are also indicated in the figures for placing the outputs of the truth table in the cells of the K-maps. The K-map in Fig. 3.10b is used in the book.

2-Bit Gray code	Adjacent row labels
0 0	A' B'
0 1	A' B
1 1	A B
1 0	A B'

	C'	C
A' B'	m_0	m_1
A' B	m_2	m_3
A B	m_6	m_7
A B'	m_4	m_5

	B' C'	B' C	B C	B C'
A'	m_0	m_1	m_3	m_2
A	m_4	m_5	m_7	m_6

(a) 2-Bit Gray code with the labels of adjacent rows of K-map

(b) 3-Variable K-map with four rows & two columns

(c) 3-Variable K-map with two rows & four columns

Fig. 3.10 Gray code and its application for K-map

3.5.2.2 K-map and Looping

A truth table with three variables is shown in Fig. 3.11a. The minterms, m_0–m_7, for the outputs of the truth table are also indicated. The K-map for the truth table is shown in Fig. 3.11b. The adjacent 1-cells of K-map are looped such that,

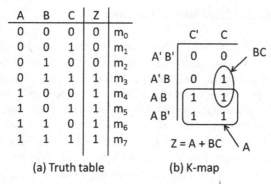

A	B	C	Z	
0	0	0	0	m_0
0	0	1	0	m_1
0	1	0	0	m_2
0	1	1	1	m_3
1	0	0	1	m_4
1	0	1	1	m_5
1	1	0	1	m_6
1	1	1	1	m_7

(a) Truth table (b) K-map

Fig. 3.11 Truth table and K-map for 3-variable function

(i)   The number of 1-cells in a loop should be 2 (pairs), 4 (quads), 8 (octets) i.e. 2^n cells.
(ii) The number of 1-cells in a loop should be large.
(iii) Overlapping of loops is permitted.
(iv) The number of loops in the K-map should be least.

Adhering to the rules of looping, there is one loop with quads and another loop with pairs. Both the loops are shown in Fig. 3.11b.

3.5.2.3 Obtaining Simplified Logic Function

The simplified logic function is obtained from the minterms of the loops with pairs and quads in Fig. 3.11b. A' and A differ by one bit in the pair loop and they are discarded. The output of the loop is BC. B', B, C' and C differ by one bit in the quad loop and they are discarded. The output of the loop is A. The simplified logic function in SOP form is:

$$Z = A + BC$$

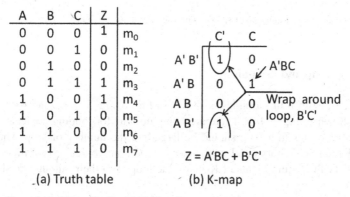

A	B	C	Z	
0	0	0	1	m_0
0	0	1	0	m_1
0	1	0	0	m_2
0	1	1	1	m_3
1	0	0	1	m_4
1	0	1	0	m_5
1	1	0	0	m_6
1	1	1	0	m_7

(a) Truth table (b) K-map

Fig. 3.12 Example 2: K-map for 3-variable function

3.5.2.4   Example 2: K-map with Three Variables

The truth table of a function with three variables and its representation in K-map are shown in Fig. 3.12a, b respectively. One looping is shown in the K-map with half of the loop is around m_0 and the other half of the loop is around m_4. It could be imagined that the looping would be complete if the K-map is wrapped around. Vertical and horizontal wrap around loops are permitted. The output of the vertical wrap around loop is $B'C'$ discarding the variables A' and A as they differ by one bit.

There may be 1-cells in the K-map which could not be looped. Such 1-cells should be included in the simplified logic function of the K-map. The 1 at the location of the minterm, m_3, and hence it is included in the logic function. The simplified logic function in SOP form is,

$$Z = A'BC + B'C'$$

3.5.3   K-map with Four Variables

The truth table for a function with four variables has 16 outputs. The general representation of the truth table in K-map with four rows and four columns is shown in Fig. 3.13. The ordering of the variables that identify the rows and columns follow Gray code. The minterms of the truth table are also indicated in the K-map for placing the outputs of the truth table.

	C' D'	C' D	C D	C D'
A' B'	m_0	m_1	m_3	m_2
A' B	m_4	m_5	m_7	m_6
A B	m_{12}	m_{13}	m_{15}	m_{14}
A B'	m_8	m_9	m_{11}	m_{10}

Fig. 3.13 General representation of K-map with four variables

3.5.3.1   K-map and Looping

A truth table with four variables is shown in Fig. 3.14a. The K-map for the truth table is shown in Fig. 3.14b. There are many options for looping adjacent 1-cells. The number of 1 s in a loop should be large and the number of loops should be least in K-map. One horizontal wrap around quad, one vertical quad and one octet loops satisfy the requirements of K-map. The loops and their outputs are shown in Fig. 3.14b.

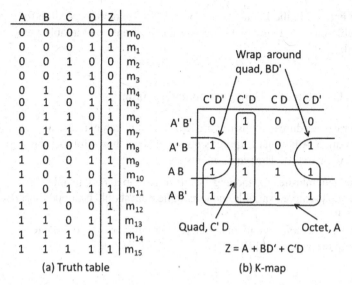

A	B	C	D	Z	
0	0	0	0	0	m_0
0	0	0	1	1	m_1
0	0	1	0	0	m_2
0	0	1	1	0	m_3
0	1	0	0	1	m_4
0	1	0	1	1	m_5
0	1	1	0	1	m_6
0	1	1	1	0	m_7
1	0	0	0	1	m_8
1	0	0	1	1	m_9
1	0	1	0	1	m_{10}
1	0	1	1	1	m_{11}
1	1	0	0	1	m_{12}
1	1	0	1	1	m_{13}
1	1	1	0	1	m_{14}
1	1	1	1	1	m_{15}

(a) Truth table

(b) K-map

Fig. 3.14 Truth table and K-map for 4-variable function

The output of the wrap around quad is BD' discarding the variables A, A′, C and C′ as they differ by one bit in the minterms of the loop. The output of the vertical quad is $C'D$ discarding the variables A, A′, B and B′ as they differ by one bit. The output of the octet is A discarding the variables B, B′, C, C′, D and D′ as they differ by one bit. The simplified logic function in SOP form is:

$$Z = A + BD' + C'D$$

3.5.4 Variable-Entered K-map

One or more variables of logic function are placed in the cells of K-map for reducing the complexity of the K-map. For example, K-map for a function with five variables requires 32 cells in the K-map. If one of the five variables could be entered into the K-map, simplified logic function could be obtained using a K-map with 16 cells, reducing the size of the K-map. The variable that is placed in the cells of K-map is called map-entered variable and the K-map is termed as variable-entered map. The map-entered variable could be the complement of the variable also.

Theoretically, if m variables out of the n variables of the function could be entered in the cells of K-map, the number of cells in the variable-entered K-map would be $2^{(n-m)}$; n should be greater than m. In practice, the number of variables (m) that could be entered in the K-map is one or two. The general procedure for construct-ing variable-entered K-map and looping the map-entered variable with 1-cells is explained.

3.5.4.1 Procedure for Construction and Looping

The general procedure for constructing K-map is applicable for constructing variable-entered K-map. Additional requirements should be adhered for looping map-entered variables with 1-cells. The requirements are:

(i)   The general rules for looping 1-cells in K-map are applicable.
(ii) The map-entered variable and its complement should be looped separately with 1-cells.
(iii) The 1-cell adjacent to map-entered variable or its complement should be completely covered [2].

3.5.4.2 Understanding Completely Covered 1-Cell

The 1-cell adjacent to map-entered variable or its complement is treated as the logical sum of the map-entered variable and its complement. If C is a map-entered variable, the 1-cell adjacent to C or C' is treated as C + C'. Loops are formed with C and C' for covering the 1-cell completely. Four examples are provided to illustrate completely covered 1-cell.

Example 1 Variable-entered K-map with the map-entered variable, C, is shown in Fig. 3.15a. The adjacent 1-cell in the location, AB', is treated as (C + C'). The map-entered variable, C, is looped with the C of the adjacent 1-cell. Another loop is formed with C and C' of the 1-cell for covering the 1-cell completely. The loops, the outputs of the loops and the simplified logic function are shown in Fig. 3.15b.

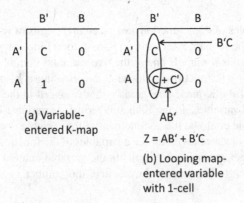

(a) Variable-entered K-map

$Z = AB' + B'C$

(b) Looping map-entered variable with 1-cell

Fig. 3.15 Example 1: looping map-entered variable with 1-cell

Example 2 Variable-entered K-map with the complement of map-entered variable, C', is shown in Fig. 3.16a. The adjacent 1-cell in the location, AB', is treated as $(C' + C)$. The complement of map-entered variable, C', is looped with the C' of the adjacent 1-cell. Another loop is formed with C' and C of the 1-cell for covering the 1-cell completely. The loops, the outputs of the loops and the simplified logic function are shown in Fig. 3.16b.

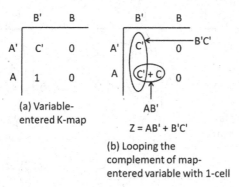

(a) Variable-
entered K-map

$Z = AB' + B'C'$

(b) Looping the
complement of map-
entered variable with 1-cell

Fig. 3.16 Example 2: looping map-entered variable with 1-cell

Example 3 Variable-entered K-map with the map-entered variable, C, and its complement, C', is shown in Fig. 3.17a. The adjacent 1-cell in the location, AB', is treated as $(C + C')$. The map-entered variable, C, is looped with the C of the adjacent 1-cell. Another loop is formed with the complement of the variable, C', and C' of the adjacent 1-cell for covering the 1-cell completely. The loops, the outputs of the loops and the simplified logic function are shown in Fig. 3.17b.

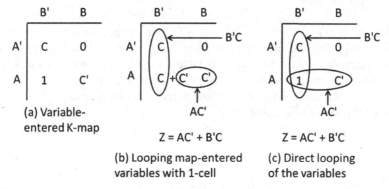

(a) Variable-
entered K-map

$Z = AC' + B'C$

(b) Looping map-entered
variables with 1-cell

$Z = AC' + B'C$

(c) Direct looping
of the variables

Fig. 3.17 Example 3: looping map-entered variable with 1-cell

As a rule of thumb, if 1-cell is looped with its adjacent map-entered variable and with its complement as in Fig. 3.17c, the 1-cell is completely covered. The simplified logic function could be obtained directly from the outputs of the loops and the function is indicated in Fig. 3.17c.

Example 4 Variable-entered K-map with the map-entered variable, C, is shown in Fig. 3.18a. The adjacent 1-cell in the location, $A'B'$, is treated as $(C + C')$. The map-entered variable, C, is looped with the C of the adjacent 1-cell. Another loop is formed with the 1-cell at $A'B'$ (C and C') and the 1-cell at AB' for covering the 1-cell completely. The loops, the outputs of the loops and the simplified logic function are shown in Fig. 3.18b.

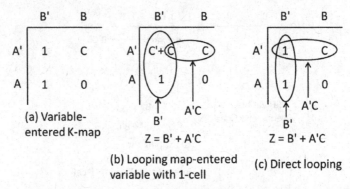

Fig. 3.18   Example 4: looping map-entered variable with 1-cell

Alternatively, looping could be done as in Fig. 3.18c for completely covering the 1-cell adjacent to the map-entered variable. The loops, the outputs of the loops and the simplified logic function are shown in the figure. Reference [2] could be referred for additional information on variable-entered K-map with more examples.

3.5.4.3   Illustration

The truth table in Fig. 3.14a for a function with four variables (n = 4) is reproduced as Fig. 3.19a to keep the illustration simple. The K-map would require 16 cells to accommodate the 16 outputs of the truth table. One (m = 1) of the four variables is selected as map-entered variable. The variable-entered K-map requires $2^{(n-m)}$ i.e. 2^3 cells. The construction of variable-entered K-map with eight cells for the function with four variables is illustrated.

A	B	C	D	Z	
0	0	0	0	0	m_0
0	0	0	1	1	m_1
0	0	1	0	0	m_2
0	0	1	1	0	m_3
0	1	0	0	1	m_4
0	1	0	1	1	m_5
0	1	1	0	1	m_6
0	1	1	1	0	m_7
1	0	0	0	1	m_8
1	0	0	1	1	m_9
1	0	1	0	1	m_{10}
1	0	1	1	1	m_{11}
1	1	0	0	1	m_{12}
1	1	0	1	1	m_{13}
1	1	1	0	1	m_{14}
1	1	1	1	1	m_{15}

(a) Truth table

A	B	C	Z	
0	0	0	D	m_0
0	0	1	0	m_1
0	1	0	1	m_2
0	1	1	D'	m_3
1	0	0	1	m_4
1	0	1	1	m_5
1	1	0	1	m_6
1	1	1	1	m_7

(b) Truth table with map-entered variable

Fig. 3.19 Truth table with map-entered variable

3.5.4.4 Truth Table with Map-Entered Variable

The truth table of the function in Fig. 3.19a has 16 minterms (m_0 to m_{15}). It is reduced to the next lower level truth table with 8 minterms (m0–m7) by selecting the variable, D, of the function as map variable. The procedure for deriving the lower level truth table is:

(i) Consider the first pair of combinations of the inputs of the truth table:

A	B	C	D	Z	Minterm
0	0	0	0	0	m_0
0	0	0	1	1	m_1

$Z = 0$ i.e. D for m_0. $Z = 1$ i.e. D for m_1. Hence, Z could be replaced by the map variable, D for m_0 and m_1. The truth table reduces to:

A	B	C	Z	Minterm
0	0	0	D	m_0

(ii) Consider the second pair of combinations of the inputs:

A	B	C	D	Z	Minterm
0	0	1	0	0	m_2
0	0	1	1	0	m_3

$Z = 0$ when $D = 0$ for m_2. $Z = 0$ when $D = 1$ for m_3. Hence, Z could be replaced by 0 for m_2 and m_3. The truth table reduces to:

A	B	C	Z	Minterm
0	0	1	0	m_1

(iii) Consider the third pair of combinations of the inputs:

A	B	C	D	Z	Minterm
0	1	0	0	1	m_4
0	1	0	1	1	m_5

$Z = 1$ when $D = 0$ for m_4. $Z = 1$ when $D = 1$ for m_5. Hence, Z could be replaced by 1 for m_4 and m_5. The truth table reduces to:

A	B	C	Z	Minterm
0	1	0	1	m_2

(iv) Consider the fourth pair of combinations of the inputs:

A	B	C	D	Z	Minterm
0	1	1	0	1	m_6
0	1	1	1	0	m_7

$Z = 1$ i.e. D' for m_6. $Z = 0$ i.e. D' for $m_{\backslash 7}$. Hence, Z could be replaced by the map variable, D' for m_6 and m_7. The truth table reduces to:

A	B	C	Z	Minterm
0	1	1	D'	m_3

The process is continued in pairs of combinations of the inputs of the truth table up to m14 and m15. The reduced truth table with map-entered variables is shown in Fig. 3.19b. The outputs (Z) of the lower level truth table contain D, D', 1 and 0.

3.5.4.5 Construction of K-map and Looping

Variable-entered K-map with the map-entered variable (D) and its complement (D') is shown in Fig. 3.20a. Three loops are shown for the K-map in Fig. 3.20b. The vertical quad loop contains the map-entered variable, D. Another quad loop contains the complement of the map-entered variable, D', and the third quad loop contains 1-cells.

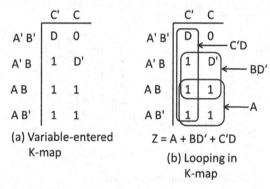

(a) Variable-entered K-map

$Z = A + BD' + C'D$

(b) Looping in K-map

Fig. 3.20 Variable-entered K-map and looping

The 1-cell in the location A'BC' is adjacent to both D and D'. It is looped by the quad with D and by another quad with D', covering the 1-cell in the location A'BC' completely.

There is another 1-cell in the location ABC and it is adjacent to D'. It is looped by the quad with D' and by the quad with 1-cells, covering the 1-cell in the location ABC completely. The outputs of the loops and the simplified logic function are shown in Fig. 3.20b. The logic function agrees with that obtained using K-map with 16 cells for the truth table in Sec. 3.5.3.1.

3.5.5 Don't-Care Conditions

In the design of digital systems, there are a few input logic conditions that never occur and hence the outputs for the input logic conditions also never occur. Such logic conditions are called don't-care conditions. Don't-cares are an important concept in logic synthesis and are frequently used for the optimization of logic circuits [3]. Don't-care conditions are shown as X in the outputs of truth table.

Don't-care conditions (X) in K-map could be treated as 1 or 0 to our advantage for obtaining simplified logic function. Hence, X can be looped with 1 s to obtain logic function in SOP form or it can be looped with 0 s to obtain logic function in POS form. Obtaining simplified logic function in SOP form is illustrated.

3.5.5.1   Simplified Function from K-map

The truth table of a function with three variables and its representation in K-map are shown in Fig. 3.21a, b respectively. The rules of looping X and 1 are same as those for looping 1 s in K-map except that looping X alone is not permitted. Two quads with 1 and X, the outputs of the quads and the simplified logic function are shown in Fig. 3.21b.

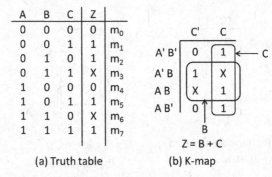

(a) Truth table   (b) K-map

Fig. 3.21 Truth table and K-map with don't-care conditions

3.5.6   Logic Function in POS Form

The general procedure for obtaining simplified logic function in SOP form is applicable for obtaining the function in POS form except for:

(i) The locations of placing the outputs of truth table in the cells of K-map are indicated by maxterms.

(ii) The sum forms of the variables of logic function are used to label the rows and columns of K-map. The simplified logic function could be obtained directly from the K-map. Gray code is used for labeling the cells.

(iii) The variables having logic level of 1 are complemented.

Labelling the rows and columns of K-map and the locations of placing the outputs of truth table in the cells of the K-map are shown for functions with two, three and four variables in Fig. 3.22a–c respectively.

(a) 2-variable function

	B	B'
A	M_0	M_1
A'	M_2	M_3

(b) 3-variable function

	C	C'
A + B	M_0	M_1
A + B'	M_2	M_3
A' + B'	M_6	M_7
A' + B	M_4	M_5

(c) 4-variable function

	C + D	C + D'	C' + D'	C' + D
A + B	M_0	M_1	M_3	M_2
A + B'	M_4	M_5	M_7	M_6
A' + B'	M_{12}	M_{13}	M_{15}	M_{14}
A' + B	M_8	M_9	M_{11}	M_{10}

Fig. 3.22 K-maps for logic functions in POS form

3.5.6.1 Logic Function from 3-Variable K-map

The truth table of logic function with three variables is shown in Fig. 3.23a. The maxterms, M_0–M_7, for the outputs of the truth table are also indicated. The K-map for the truth table is shown in Fig. 3.23b. Three pairs of loops with 0s are shown in the figure. The output of each loop is in sum-form, discarding the variables that differ by one bit in maxterms of the loop. The simplified logic function in POS form is:

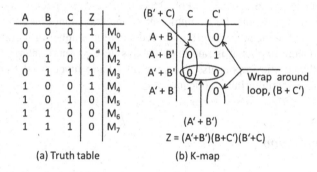

Fig. 3.23 Logic function in POS form from 3-variable K-map

$$Z = (A' + B')(B + C')(B' + C)$$

3.6 Quine-McCluskey Method

In 1956, Edward McCluskey created the Quine-McCluskey logic minimization procedure as a doctoral student at MIT, USA: the first systematic method to obtain a minimized two-level logic representation of a digital circuit, marking the beginning of electronic design automation [4]. Quine-McCluskey method is useful for obtaining the simplified logic function of truth table with any number of variables in SOP form and the method could be computerized. The minterms having 1 as outputs of the truth table are grouped and they are combined with the minterms of the adjacent group. The combining process is continued successively with adjacent groups for obtaining simplified logic function. The application of the method is illustrated after defining the terms associated with the method.

3.6.1 Definition of Terms

Literals, implicants, prime implicants and essential prime implicants are used in the application of Quine-McCluskey method. Literal is a generic term representing the variable or its complement of logic function. If A, B and C are the variables of logic function, A, B, C, A′, B′ and C′ are the literals of the function.

The minterm having the output, 1, is termed as implicant. For example, if truth table of logic function shows 1 as outputs for three combinations of inputs, the truth table is said to contain three implicants.

The implicants of adjacent groups are combined in Quine-McCluskey method. Combining the implicants with the implicants of adjacent group reduces the number of literals in the implicants. Prime implicant is one that cannot be combined with another implicant in the adjacent group for reducing the number of literals further.

Essential prime implicant is one that contains at least one minterm that is not shared by other prime implicants. The simplified logic function of truth table is obtained from essential prime implicants.

3.6.2 Illustration

The application of Quine-McCluskey method is illustrated for the truth table shown in Fig. 3.24. The truth table contains twelve minterms with 1 as outputs i.e. twelve implicants. The implicants are successively combined to identify prime implicants. Essential prime implicants are derived from the prime implicants and they are expressed in SOP form, providing the simplified logic function.

A	B	C	D	Z	
0	0	0	0	0	m_0
0	0	0	1	1	m_1
0	0	1	0	0	m_2
0	0	1	1	0	m_3
0	1	0	0	1	m_4
0	1	0	1	1	m_5
0	1	1	0	1	m_6
0	1	1	1	0	m_7
1	0	0	0	1	m_8
1	0	0	1	1	m_9
1	0	1	0	1	m_{10}
1	0	1	1	1	m_{11}
1	1	0	0	1	m_{12}
1	1	0	1	1	m_{13}
1	1	1	0	1	m_{14}
1	1	1	1	1	m_{15}

Fig. 3.24 Truth table for 4-variable function

3.6.2.1 Combining Implicants

Step-1:

(i) Group the implicants of truth table with zero 1s, one 1s, two 1s and so on in the input combinations as applicable to the truth table.

Examining the truth table in Fig. 3.24, there are three implicants with one 1s, five implicants with two 1s, three implicants with three 1s and one implicant with one 1s. They are grouped and the implicants with their IDs are shown in Table 3.1 as Stage-1. The IDs of minterms are assigned to the implicants.

Step-2: Implicants in G1, Stage-1

(i) Begin with the implicant #1, 0001.

- Search in the adjacent group i.e. in G2, stage-1 for an implicant that differs by one bit from the implicant, 0001. The implicant #5, 0101, differs by one bit.
- Combine the two implicants. The combined implicant is identified as (1,5) and it is obtained by replacing the bit that differs by underscore.
  $0001 + 0101 = 0\_01$ i.e. $A'B'C'D + A'BC'D = A'C'D$ i.e. $A'\_C'D$
  However, the combined output is maintained with the combination of 0 and 1 and replacing the discarded variable with underscore until the completion of the Quine-McCluskey method.
- Record the combined implicant, (1,5), i.e. 0_01, in G1 of stage-2 as shown in Table 3.1.

Table 3.1 Implicants of truth table and combined implicants in stage-2

Stage-1: implicants			Stage-2: combined implicants			Stage-3: combined implicants		
Group	ABCD	ID	Group	ABCD	ID	Group	ABCD	ID
G1	0001 ✓	1	G1	0_01 ✓	(1,5)	G1	_ _01	(1,5,9,13)
	0100 ✓	4		_001 ✓	(1,9)		_ _01	(1,9,5,13)
	1000 ✓	8		010_ ✓	(4,5)		_10_	(4,5,12,13)
G2	0101 ✓	5		01_0 ✓	(4,6)		_1_0	(4,6,12,14)
	0110 ✓	6		_100 ✓	(4,12)		_10_	(4,12,5,13)
	1001 ✓	9		100_ ✓	(8,9)		_1_0	(4,12,6,14)
	1010 ✓	10		10_0 ✓	(8,10)		10_ _ ✓	(8,9,10,11)
	1100 ✓	12		1_00 ✓	(8,12)		1_0_ ✓	(8,9,12,13)
G3	1011 ✓	11	G2	_101 ✓	(5,13)		10_ _ ✓	(8,10,9,11)
	1101 ✓	13		_110 ✓	(6,14)		1_ _0 ✓	(8,10,12,14)
	1110 ✓	14		10_1 ✓	(9,11)		1_0_ ✓	(8,10,9,13)
G4	1111 ✓	15		1_01 ✓	(9,13)		1_ _0 ✓	(8,12,10,14)
				101_ ✓	(10,11)	G2	1_ _1 ✓	(9,11,13,15)
				1_10 ✓	(10,14)		1_ _1 ✓	(9,13,11,15)
				110_ ✓	(12,13)		1_1_ ✓	(10,11,14,15)
				11_0 ✓	(12,14)		1_1_ ✓	(10,14,11,15)
			G3	1_11 ✓	(11,15)		11_ _ ✓	(12,13,14,15)
				11_1 ✓	(13,15)		11_ _ ✓	(12,14,13,15)
				111_ ✓	(14,15)			

- Tick both the implicants for having completed the combining operation. Do not tick the implicants if they cannot be combined.
- Continue search in G2 for additional implicants that differ by one bit from the implicant #1, 0001. The implicant #9, 1001, differs by one bit. Combine the two implicants. Record the combined implicant #(1,9), _001, in G1 of stage-2. Tick both the implicants for having completed the combining operation.

(ii) Perform the above operations for all the implicants in G1 of Stage-1 and record the combined implicants in G1 of stage-2.

Step-3: Implicants in G2 and G3, Stage-1

(i) Perform the operations in step-2 for the implicants in G2 by combining with the implicants in G3 of stage-1 and record the combined implicants in G2 of stage-2.

(ii) Perform the operations in step-2 for the implicants in G3 by combining with the implicants in G4 of stage-1 and record the combined implicants in G3 of stage-2.

Step-4: Implicants in G1, G2 and G3, Stage-2

(i) Perform the operations in step-2 for the implicants in G1 by combining with the implicants in G2 of stage-2 and record the combined implicants in G1 of stage-3 as shown in Table 3.1.

(ii) Perform the operations in step-2 for the implicants in G2 by combining with the implicants in G3 of stage-2 and record the combined implicants in G2 of stage-3.

Step-5: Implicants in G1 and G2, Stage-3

(i) Perform the operations in step-2 for the implicants in G1 by combining with the implicants in G2 of stage-3 and record the combined implicants in G1 of stage-4 as shown in Table 3.2.

(ii) The combined implicants having the IDs, (8,9,10,11) and (8,10,9,11) are ticked as the IDs are same. Other combined implicants having the same IDs in stage-3 are also ticked.

(iii) Combining the implicants in G1, stage-4 is no more possible. The next step is to identify the prime implicants.

Table 3.2 Combined implicants in stage-3 and stage-4

Stage-3: combined implicants			Stage-4: combined implicants		
Group	ABCD	ID	Group	ABCD	ID
G1	_ _01	(1,5,9,13)	G1	1_ _ _	(8,9,10,11,12,13,14,15)
	_ _01	(1,9,5,13)		1_ _ _	(8,9,12,13,10,11,14,15)
	_10_	(4,5,12,13)		1_ _ _	(8,10,12,14,9,11,13,15)
	_1_0	(4,6,12,14)			
	_10_	(4,12,5,13)			
	_1_0	(4,12,6,14)			
	10_ _ √	(8,9,10,11)			
	1_0_ √	(8,9,12,13)			
	10_ _ √	(8,10,9,11)			
	1_ _0 √	(8,10,12,14)			
	1_0_ √	(8,10,9,13)			
	1_ _0 √	(8,12,10,14)			
G2	1_ _1 √	(9,11,13,15)			
	1_ _1 √	(9,13,11,15)			
	1_1_ √	(10,11,14,15)			
	1_1_ √	(10,14,11,15)			
	11_ _ √	(12,13,14,15)			
	11_ _ √	(12,14,13,15)			

Table 3.3 Prime implicants

Stage	Prime implicants		
	0 and 1 form ABCD	Literal form	ID from Table 3.2
Stage-3	_ _ 01	C'D	(1,5,9,13)
	_10_	BC'	(4,5,12,13)
	_1_0	BD'	(4,6,12,13)
Stage-4	1_ _ _	A	(8,9,10,11,12,13,14,15)

3.6.2.2 Identifying Prime Implicants

Prime implicants are those which could not be combined with the implicants of the adjacent groups of stages. The implicants for which the symbol, $\sqrt{}$, is absent are the prime implicants. The symbol is present for all the implicants in stage-2. The symbol is absent for six prime implicants in stage-3 and three prime implicants in stage-4. Omitting the prime implicants having the same IDs, the number of prime implicants reduces to three in stage-3 and one in stage-4. The prime implicants with the combination of 0 and 1 and with the corresponding literals are shown in Table 3.3.

3.6.2.3 Essential Prime Implicants and Logic Function

The prime implicants and the IDs of the implicants of the truth table (Fig. 3.24) are tabulated in Table 3.4 to identify the essential prime implicants. The identifications of prime implicants are used to tick the IDs of the implicants in Table 3.4. For example, the ID of the prime implicant, C'D, is (1,5,9,13) and ticks are placed at the locations, 1, 5, 9 and 13 for the prime implicant in Table 3.3. Similarly, ticks are placed for other prime implicants as per their IDs.

Essential prime implicant should have one implicant that is not shared by other prime implicants. C'D is an essential prime implicant as the implicant, 1, is not shared by other prime implicants. BC' is not an essential prime implicant as all its implicants are shared by other prime implicants. BD' is an essential prime implicant as the implicant as the implicant, 6, is not shared by other prime implicants. A is

Table 3.4 Matrix of Prime implicants and IDs implicants of truth table

Prime implicant	IDs of implicants of truth table											
	1	4	5	6	8	9	10	11	12	13	14	15
C'D	$\sqrt{}$		$\sqrt{}$			$\sqrt{}$				$\sqrt{}$		
BC'		$\sqrt{}$	$\sqrt{}$						$\sqrt{}$	$\sqrt{}$		
BD'		$\sqrt{}$		$\sqrt{}$					$\sqrt{}$		$\sqrt{}$	
A					$\sqrt{}$	$\sqrt{}$	$\sqrt{}$	$\sqrt{}$	$\sqrt{}$	$\sqrt{}$	$\sqrt{}$	$\sqrt{}$

also an essential prime implicant as the implicants, 8, 10, 11 and 15 are not shared by other implicants. The simplified logic function of truth table is the logical sum of essential prime implicants provided the essential prime implicants cover all the implicants of the truth table. If all the implicants of the truth table are not covered, additional prime implements are included in logic function to cover all the implicants. Examining Table 3.3, the three essential prime implicants cover all the implicants of truth table. The simplified logic function is:

$$Z = A + BD' + C'D$$

3.7 Hazards

Timing diagram shows changes in output logic state related to the inputs of the truth table in graphical form. Unwanted changes in output state might be observed defying the predicted behavior of the logic function of truth table. The unwanted changes in output logic state are called hazards. A simple example of predicted output and observed output with hazard is shown in Fig. 3.25. The duration of unwanted changes in output logic states is in the order of nanoseconds and hence, hazards are also called glitches or transients.

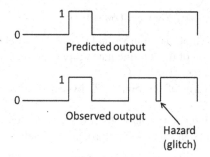

Fig. 3.25 Hazard in observed output logic states

3.7.1 Causes

Integrated circuits are available for logic gates and they consist of many transistors interconnected in monolithic form on wafers. The transistors require finite time for switching on or off due to inherent performance limitations. The finite switching time of transistors cumulatively reflects as propagation delay for a logic gate i.e. the logic gate requires finite time to respond after the application of input logic signals. The finite response time of the gate is termed as propagation delay. The propagation delay of logic gates is in the order of nanoseconds and it leads to hazards (glitches) in output logic states. Two examples are provided for understanding hazards caused by the propagation delay of logic gates.

3.7.1.1 Momentary Change to Low State

Consider the logic circuit shown in Fig. 3.26a. The circuit consists of inverter and OR gate. It is assumed that the propagation delay of both the gates is t nSec. The hazard condition in the output logic states could be simulated using software tools for any pre-defined propagation gate delays for the gates. The method of identifying the hazard condition in the output logic states using timing diagram are shown in Fig. 3.26b. The steps in the method are:

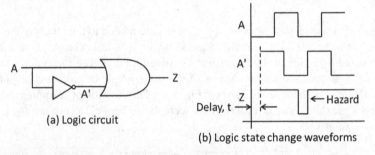

(a) Logic circuit

(b) Logic state change waveforms

Fig. 3.26 Hazard with momentary change to Low state

(i)   Draw the waveform of the input variable, A, without delay.
(ii) Draw the waveform of the intermediate output variable, A', with the propagation delay of inverter, t nSec.
(iii) Obtain the waveform of the output, Z, by performing OR operation of the waveforms of A and A'.
(iv) The glitch with momentary change to Low state is observed. The output would have remained in High state without propagation delay.

The output (Z) waveform is shown with a delay, t nSec in the figure. Actually, the waveform would actually appear with the cumulative delay of 2t nSec including the propagation delay of OR gate.

3.7.1.2 Momentary Change to High State

Consider the gate circuit shown in Fig. 3.27a. The circuit consists of inverter and AND gate. It is assumed that the propagation delay of both the gates is t nSec. The timing diagram indicating the logic state changes for A, A' and Z is drawn as per the procedure described in Sect. 3.7.1.1. It is shown in Fig. 3.27b. The glitch with momentary change to High state is also shown in the figure. The output (Z) would have remained in Low state without propagation delay.

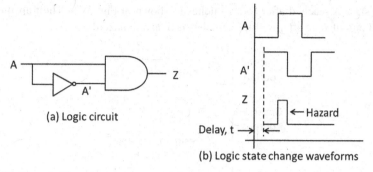

(a) Logic circuit

(b) Logic state change waveforms

Fig. 3.27 Hazard with momentary change to High state

3.7.2 Types of Hazards

Hazards are classified as static and dynamic hazards considering the nature of transients in the output of combinational logic circuits. Static hazards are further classified as static-1 and static-0 hazards.

3.7.2.1 Static-1 Hazard

The output of combinational circuit changing to Low state momentarily when it should remain High is termed as static-1 hazard. The hazard is also called SOP hazard—a glitch that occurs in otherwise steady-state 1 output signal from SOP logic [5]. The example, shown in Fig. 3.26, is static-1 hazard.

3.7.2.2 Static-0 Hazard

The output of combinational circuit changing to High state momentarily when it should remain Low is termed as static-0 hazard. The hazard is also called POS hazard—a glitch that occurs in otherwise steady-state 0 output signal from POS logic [5]. The example, shown in Fig. 3.27, is static-0 hazard.

3.7.2.3 Dynamic Hazard

The output state of combinational circuit changing many times when it should change once is termed as dynamic hazard. For example, desired output should change from 0 to 1 in single transition for the proper functioning of logic circuit; instead the High output state is reached after three glitches as shown in Fig. 3.28. The transition to desired output state after multiple glitches is dynamic hazard.

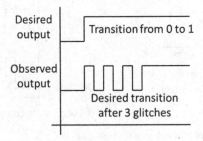

Fig. 3.28 Dynamic hazard output with multiple glitches

Dynamic hazard is not present in simple gate circuits. As the complexity of combinational gate circuit increases, static hazards in the multiple paths of complex combinational logic circuit cause dynamic hazard. If static hazards are avoided, combinational logic circuit would be free of dynamic hazards also.

3.7.3 Avoiding Hazards

Hazard free circuits are necessary for safety and critical applications. Basically, design efforts focus on removing static hazards in logic gate circuits as they would remove dynamic hazards also.

Advanced techniques are available for analyzing hazards. DILL (Digital Logic in LOTOS) can be used to specify and analyze hardware timing characteristics using ET-LOTOS (Enhanced Timed LOTOS), a timed extension of the ISO standard formal language LOTOS (Language of Temporal Ordering Specification) and delays are placed in series with the functionality to provide delay for each output based on the analysis [5]. Glitch reduction optimization algorithm sets the don't-cares of logic function in such a way that reduces the amount of glitching and this process is performed after placement and routing, using timing simulation data to guide the algorithm [3].

Balancing propagation delay is used by using delay circuits for eliminating hazards The datasheets of ICs for combinational digital hardware specify 'balanced propagation delays' for avoiding hazards. Including a redundant prime implicant of truth table is another simple method for avoiding static hazards.

3.7.3.1 Redundant Implicant to Avoid Static-1 Hazard

The truth table of logic function with three variables, the K-map of the truth table and the implementation of the simplified logic function in SOP form are shown in Fig. 3.29. One inverter and two AND gates are used for the implementation. Let C',

X and Y be the intermediate outputs of the gates. Z is the final output of the gate circuit. It is assumed that the propagation delay of each gate is t nSec.

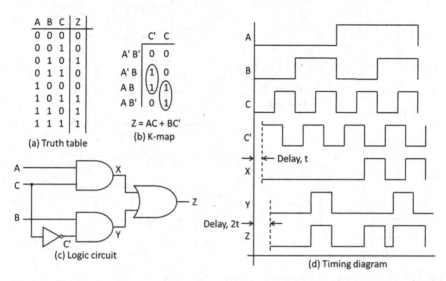

A B C	Z
0 0 0	0
0 0 1	0
0 1 0	1
0 1 1	0
1 0 0	0
1 0 1	1
1 1 0	1
1 1 1	1

(a) Truth table

Z = AC + BC'

(b) K-map

(c) Logic circuit

(d) Timing diagram

Fig. 3.29 Static-1 hazard in 3-variable logic circuit

The timing diagram of the gate circuit is shown in Fig. 3.29d. The waveforms of input, intermediate and final output logic signals are shown in the timing diagram. The waveforms of C′ and X are shown with a delay of t nSec. The waveforms of Y and Z are shown with the cumulative delay of 2t nSec. The final output waveform shows static-1 hazard, caused by the propagation delay of the gates. Actually, the output waveform (Z) with the hazard would appear with the cumulative delay 3t nSec.

The static-1 hazard of the logic circuit in 3.29c could be eliminated by including additional prime implicant in the logic function. The redundant prime implicant, AB, is shown by the horizontal loop in Fig. 3.30b. The modified logic diagram is shown in Fig. 3.30c. The intermediate output of the prime implicant is H. The waveforms of the input, intermediate and output logic signals are shown in the timing diagram with appropriate delays in Fig. 3.30d. The static-1 hazard is eliminated in the final output (Z) of the gate circuit.

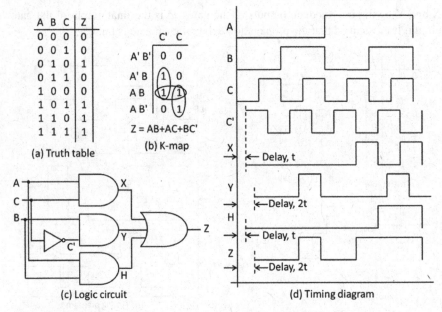

(a) Truth table

(b) K-map

(c) Logic circuit

(d) Timing diagram

Fig. 3.30 Eliminating static-1 hazard in 3-variable logic circuit

3.7.3.2 Redundant Implicant to Avoid Static-0 Hazard

The truth table of 3-variable logic function and the K-map for the truth table are shown in Fig. 3.31a, b. The simplified logic function in POS form for the K-map in 3.31b is $[(A + C) (B + C')]$. The logic function is implemented using inverter, OR and AND gates. Assume that the propagation delay of each gate is t nSec. Timing diagram analysis indicates static-0 hazard at the final output (Z) of the gate circuit. The diagram is not shown in the figure.

A	B	C	Z
0	0	0	0
0	0	1	0
0	1	0	1
0	1	1	0
1	0	0	0
1	0	1	1
1	1	0	1
1	1	1	1

(a) Truth table

	C'	C
A+B	0	0
A+B'	1	0
A'+B'	1	1
A'+B	0	1

Z = (A+C) (B+C')

Implementing the logic function results in static-0 hazard

(b) K-map: Essential prime implicants

	C'	C
A+B	0	0
A+B'	1	0
A'+B'	1	1
A'+B	0	1

Z = (A+B) (A+C) (B+C')

Implementing the logic function removes static-0 hazard

(c) K-map: Essential prime implicants & additional prime implicant, (A+B)

Fig. 3.31 Eliminating static-0 hazard in 3-variable logic circuit

The hazard could be eliminated by including redundant prime implicant, (A + B), in the logic function of the gate circuit. The redundant implicant is shown by the horizontal loop in Fig. 3.31c. The logic function becomes [(A + B) (A + C) (B + C')]. Timing analysis for the modified gate circuit would indicate the elimination of the static-0 hazard at the final output (Z) of the gate circuit.

References

1. Doran R (2007) The gray code. J Univ Comput Science 13(11):1573–1597
2. Givone D (2003) Digital principles and design. McGraw-Hill Higher Education
3. Shum W, Anderson J (2007) FPGA glitch power analysis and reduction. In: ACM international symposium on low power electronics and design, IEEE
4. Flynn M, Mitra S (2017) Edward J. McCluskey 1929–2016. IEEE Des Test
5. He J, Turner KJ (2001) Specifying hardware timing with ET-LOTOS. In: Proceedings of 11th conference on Correct Hardware Design and Verification Methods (CHARME 2001), Lecture notes in computer science, vol 2144. Springer, pp 161–166

Chapter 4
Combinational Logic Devices

Abstract Combinational logic devices are readily available as standard ICs for performing complex digital functions and they are hazard free. The operation of multiplexers, demultiplexers, decoders, encoders, parity generators, parity checkers and comparators are presented. The application of the devices is also presented.

4.1 Introduction

Due to phenomenal growth in the scale of integration, the density of gates in a chip has increased considerably. Combinational logic devices are available as standard ICs for performing complex digital functions required for data transmission, code conversion and arithmetic operations. The ICs are reliable and hazard free. They could be used readily for designing digital signal processing (DSP) circuits. However, it is necessary to understand the functioning of general purpose logic devices that are used in large number of DSP applications. The general purpose logic devices are similar to the analog devices such as op-amps and 3-terminal linear voltage regulators.

Multiplexers, demultiplexers, decoders, encoders, parity generators, parity checkers, comparators, adders and subtractors are some of the commonly used general purpose combinational logic devices. The operation of the devices except adders and subtractors is explained in this chapter. Adders and subtractors are explained in Chap. 6.

4.2 Multiplexers

A multiplexer is a device that has input signals from multiple data sources. The device outputs one of the input signals selected by a set of control signals. For example, if the output digital signals of eight logic circuits (data sources) need to be connected to the input of microcontroller for further operations, the eight digital signals are

© Springer Nature Switzerland AG 2020
D. Natarajan, *Fundamentals of Digital Electronics*,
Lecture Notes in Electrical Engineering 623,
https://doi.org/10.1007/978-3-030-36196-9_4

routed through a multiplexer. The required data source signal is selected by the control signals of the multiplexer. The selected signal is transmitted to the input of the microcontroller using the common transmission line between the multiplexer and the microcontroller. As multiplexer performs parallel-to-serial conversion of data, it is used in long distance communication for avoiding expensive parallel data transmission lines.

The functioning of a multiplexer could be equated to a rotary wafer switch in analog circuits. Rotary switch has a pole contact terminal and many radial contact terminals. The electrical signals of many analog circuits are connected to the radial terminals of the switch. The switch is manually rotated and the pole of the switch selects the required signal for further processing. The selected signal is routed through a common line for further processing such as measuring the voltage level of the signal.

4.2.1　Operation

Multiplexer is also called data selector as it selects one of the output from many data sources. The control signals for selecting the desired data source are select inputs to multiplexer. The number of select inputs and the number of input data sources are related. If S is the number of select inputs of multiplexer, the number of data sources that could be connected to the multiplexer should be less than or equal to 2^S. If $S = 1$, two data sources could be connected to multiplexer and the multiplexer is called 2:1 MUX. If $S = 2$, four data sources could be connected to multiplexer and the multiplexer is called 4:1 MUX. Input data sources are numbered as D_0, D_1, D_2 and so on. Select inputs are identified as A, B, C and so on.

4.2.1.1　Operation of 2:1 MUX

The block diagram of 2:1 MUX with data inputs (D_0 and D_1) from logic circuits, select input (A) and MUX output (Z) is shown in Fig. 4.1a. Timing diagram of the MUX is shown in Fig. 4.1b. The output of the MUX for the assumed select input and data signals is shown in the diagram. The truth table of the MUX is shown in Fig. 4.1c.

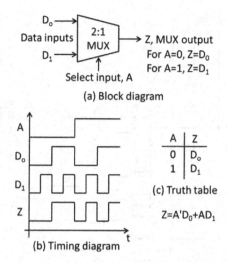

Fig. 4.1 Functioning of 2:1 MUX

When the logic status of the select input is Low, the logic signal from the data source, D_0 is selected and the output of the multiplexer is D_0. When the logic status of the select input is High, the logic signal from the data source, D_1 is selected and the output of the multiplexer is D_1. The logic function of 2-input multiplexer, Z, is given by:

$$Z = A'D_0 + AD_1$$

4.2.1.2 Operation of 4:1 MUX

The logic circuit diagram of 4:1 MUX is shown in Fig. 4.2. The select inputs are A and B. The data inputs are D_0, D_1, D_2 and D_3. They are connected to the four AND gates of the MUX. The OR gate outputs the selected data signal. Let Z be the output of the multiplexer.

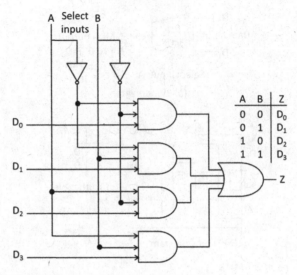

Fig. 4.2 Logic circuit diagram of 4:1 MUX

The select input circuit has two inverters and the circuit outputs four combinations of select inputs, The combinations of inputs to the four AND gates are $A'B'$ ($A = 0$ & $B = 0$), $A'B$ ($A = 0$ & $B = 1$), AB' ($A = 1$ & $B = 0$) and AB ($A = 1$ & $B = 1$). One of the AND gate outputs goes High for each input combination of the select inputs. The multiplexer outputs the data signal connected to the AND gate. The output, Z, is D_0 for the select inputs, $AB = 00$; $Z = D_1$ for $AB = 01$ and so on. The outputs of the multiplexer for the select inputs are also shown in Fig. 4.2. The logic function of 4-input multiplexer, Z, is given by:

$$Z = A'B'D_0 + A'BD_1 + AB'D_2 + ABD_3$$

4.2.1.3 Multiplexer ICs

Multiplexer ICs are available in various configurations. There is no need to design and construct multiplexers using basic gates. For example, multiplexer IC, 74150, is available and it is used for selecting one-of-sixteen data sources (16:1 MUX). Multiplexer IC, 74151, is 8:1 MUX. Standard ICs have strobe input. Active Low signal to strobe input enables the operation of the ICs. The datasheets of device manufactures could be referred for connection diagrams and truth tables.

4.2.2 Implementing 3-Variable Truth Table Using 8:1 MUX

The standard ICs of multiplexers could be used for implementing the logic functions of truth tables directly in SOP form without the need to apply Boolean algebra or K-map. Multiplexer is called universal logic circuit as the logic function of truth table with more number of variables could be realized. The implementation of logic function is illustrated for truth table with three variables using 8:1 MUX. The method could be extended for implementing the logic function of truth table with four or more number of variables.

4.2.2.1 Generating Logic Function

The truth table of 3-variable logic function is shown in Fig. 4.3a. A multiplexer having 2^3 i.e. 8 inputs is required for implementing the logic function of truth table with three variables. Assume that the 8:1 MUX IC, 74150, is used for implementing the logic function of the truth table. D_0 to D_7 are the data inputs to the IC 74150. Linking the data inputs of the IC to the minterms of the truth table is shown in the figure.

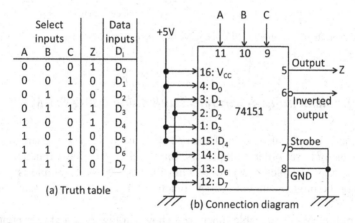

Fig. 4.3 Generating 3-variable function using MUX

The data inputs of the IC that are linked to the minterms having 1 as output are connected to V_{CC} (+5 V). The data inputs of the IC that are linked to the minterms having 0 as output are grounded. The connection diagram is shown in Fig. 4.3b. The active Low strobe input of the IC is also grounded for the operation of the IC. The generated logic function (Z) is available at the output pin #5 of the IC.

4.2.2.2   Logic Function at Inverted Output of IC

The logic function of the truth table could be generated at the inverted output pin #6 of the IC, 74151. The outputs of the truth table are inverted to obtain Z'. The complemented outputs having 1 as output are connected to V_{CC} (+5 V) and the outputs having 0 are grounded. The truth table with Z' and the connection diagram of the 8:1 multiplexer are shown in Fig. 4.4. The output at the pin #6 of the multiplexer is the logic function (Z) as both the input and output of the multiplexer are inverted.

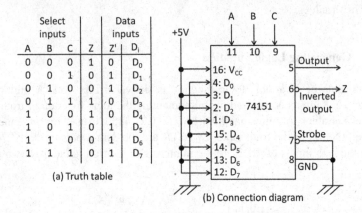

Fig. 4.4 Generating 3-variable function using MUX with inverted input data

4.2.3   Implementing 4-Variable Truth Table Using 8:1 MUX

Logic function could also be implemented for truth table with higher number of variables using lower order multiplexers. Implementing the logic function of truth table with four variables normally requires 16:1 MUX. An example is provided illustrating the implementation of logic function for truth table with four variables using 8:1 MUX.

The truth table of 4-variable function is shown in Fig. 4.5a and the variables of the truth table are A, B, C and D. The number of terms in the truth table is reduced to eight by selecting D as the map-entered variable (Sect. 3.5.4.4).

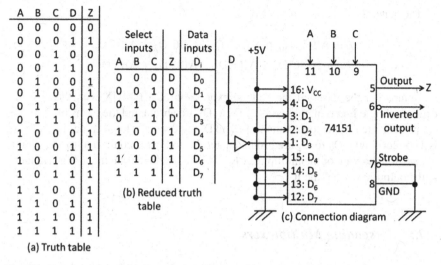

Fig. 4.5 Generating 4-variable function using 8:1 MUX

The reduced truth table with map-entered variable is shown in Fig. 4.5b. The map-entered variable, D, is considered as data input to the multiplexer. Accordingly, D is connected to the pin# 4 of the IC and its complement, D′, is connected to pin# 1. The connection to the 8:1 MUX is shown in Fig. 4.5c. The generated logic function (Z) is available at the output pin #5 of the IC.

4.2.4 Direct Implementation of Logic Function

Logic function could be directly implemented using multiplexer without preparing truth table. The logic expression of multiplexer is used for the implementation. Let Z be the output of the simplified logic function.

$$Z = A + BC$$

As the logic function contains three variables, a multiplexer having 2^3 inputs i.e. 8:1 MUX is required for implementing the function. The logic function is expanded in the form of the expression for 8:1 MUX using Boolean algebra.

$$A = AB + AB' = (ABC + ABC') + (AB'C + AB'C')$$
$$BC = ABC + A'BC$$
$$Z = (ABC + ABC') + (AB'C + AB'C') + ABC + A'BC$$
$$Z = A'BC + AB'C' + AB'C + ABC' + ABC$$

The general expression of 8:1 MUX is:

$$Z = A'B'C'D_0 + A'B'CD_1 + A'BC'D_2 + A'BCD_3$$
$$+ AB'C'D_4 + AB'CD_5 + ABC'D_6 + ABCD_7$$

Comparing the terms of the general expression of 8:1 MUX with those of the expanded logic function, the terms of D_0 to D_2 are absent and the terms of D_3 to D_7 are present. Hence, the logic values of D_0 to D_2 are 0 and the logic values of D_3 to D_7 is 1. Accordingly, D_0 to D_2 are grounded and D_3 to D_7 are connected to V_{CC} in the connection diagram of the multiplexer IC, 74151. The logic function (Z) is available at the output pin #5 of the IC.

4.2.5 Cascading Multiplexers

Higher order multiplexers are needed in digital signal processing applications. Higher order multiplexers are not readily available and they are implemented by cascading lower order multiplexers. For example, 64:1 MUX or 256:1 MUX could be implemented by using 8:1 MUX or by using 16:1 MUX.

4.2.5.1 General Expression for Multiplexers

The approach for implementing higher order multiplexer using lower order multiplexers is derived from the general expression for 2^n:1 MUX. The output of multiplexer is given by the expression,

$$Z = \sum_{i=0}^{(2^n-1)} m_i D_i$$

D_i is the data input and m_i is the minterm of the select inputs of MUX. The expression for the outputs of 2:1 and 4:1 multiplexers could be derived from the general expression.

For 2:1 MUX:

$n = 1$

Data inputs: D_0 and D_1

Select input: A

Minterms of select input: $m_0 = A'$ and $m_1 = A$

$Z = m_0 D_0 + m_1 D_1 = A'D_0 + AD_1$.

For 4:1 MUX:

$n = 2$

Data inputs: D_0, D_1, D_2 and D_3

Select inputs: A and B

Minterms of select inputs: $m_0 = A'B'$, $m_1 = A'B$, $m_2 = AB'$, and $m_3 = AB$

$Z = m_0D_0 + m_1D_1 + m_2D_2 + m_3D_3 = A'B'D_0 + A'BD_1 + AB'D_2 + ABD_3$

4.2.5.2 Implementing 8:1 MUX Using 2:1 and 4:1 MUXes

The expression for the output of 8:1 MUX is:

$n = 3$

$Z = m_0D_0 + m_1D_1 + m_2D_2 + m_3D_3 + m_4D_4 + msD_5 + m_6D_6 + m_7D_7$

$Z = A'B'C'D_0 + A'B'CD_1 + A'BC'D_2 + A'BCD_3 + AB'C''D_4 + AB'CD_5$
$\quad + ABC'D_6 + ABCD_7$

The expression is re-arranged as:

$$Z = A'B'(C'D_0 + CD_1) + A'B(C'D_2 + CD_3)$$
$$+ AB'_3(C'D_4 + CD_5) + AB(C'D_6 + CD_7)$$

$(C'D_0 + CD_1)$, $(C'D_2 + CD_3)$, $(C'D_4 + CD_5)$ and $(C'D_6 + CD_7)$ represent the outputs of four 2:1 multiplexers. The data inputs of the first multiplexer are D_0 and D_1. The data inputs of the second multiplexer are D_2 and D_3. The data inputs of the third multiplexer are D_4 and D_5. The data inputs of the fourth multiplexer are D_6 and D_7. The select input to the four multiplexers is C.

Let $I_0 = (C'D_0 + CD_1)$; $I_1 = (C'D_2 + CD_3)$; $I_2 = (C'D_4 + CD_5)$; $I_3 = (C'D_0 + CD_1)$

$$Z = A'B'I_0 + A'BI_1 + AB'I_2 + ABI_3$$

The expression for Z represents the output of 4:1 MUX. The data inputs to the 4:1 MUX are the outputs of 2:1 MUXes. The data inputs to 4:1 MUX are I_0, I_1, I_2 and I_3 The select inputs to the multiplexer are A and B. The implementation of 8:1 MUX using 2:1 and 4:1 MUXes is shown in Fig. 4.6.

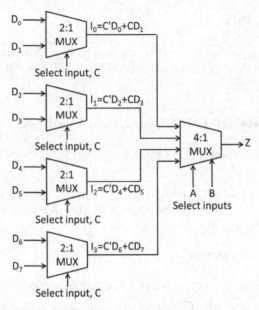

Fig. 4.6 Implementation of 8:1 MUX using 2:1 and 4:1 MUXes

Similarly, 32:1 MUX could be implemented by cascading four 8:1 MUXes at the input followed by one 8:1 MUX. Alternatively, it could be implemented by cascading two 16:1 MUXes at the input followed by one 2:1 MUX.

4.2.5.3 Cascading Standard Multiplexer ICs

Higher order multiplexers are easily implemented by cascading standard multiplexer ICs. Standard ICs have Enable input for cascading. For example, 8:1 MUX could be implemented by cascading two 4:1 MUX ICs followed by OR gate. The functional block diagram for the implementation of 8:1 MUX is shown in Fig. 4.7. One of the two 4:1 MUXes is selected by the active Low Enable input. If Enable input is Low, the multiplexer with data inputs, D_0 to D_3, is selected. If Enable input is High, the multiplexer with data inputs, D_4 to D_7, is selected. The required data source is available at the output (Z) of OR gate and it is controlled by the Select inputs, S_0 and S_1.

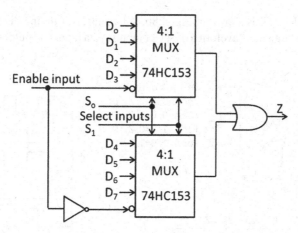

Fig. 4.7 Implementation of 8:1 MUX using 4:1 MUX ICs

4.3 Demultiplexers

A demultiplexer has one data input, select inputs and many output lines. It routes the input data signal to one of the output lines. The selection of the output line is controlled by select input signals. Basically, a demultiplexer performs the reverse operation of multiplexer. It performs serial-to-parallel conversion of data.

Demultiplexer could also be equated to a rotary wafer switch. For example, assume that DC power supply is connected to the pole terminal of the wafer switch and many analog circuits are connected to the radial terminals of the switch. The switch is manually rotated and the pole of the switch connects the DC supply to the required circuit.

The number of select inputs of demultiplexer is related to number data output lines (or simply outputs) of the demultiplexer. If S is the number of select inputs of demultiplexer, the number of data outputs that are available for distribution from the demultiplexer is less than or equal to 2^S. If S = 1, two data outputs could be realized from the demultiplexer. The demultiplexer is called 1:2 DEMUX. If S = 2, four or less data outputs could be realized from the demultiplexer. The multiplexer is called 1:4 DEMUX. Data outputs are numbered as Z_0, Z_1, Z_2 and so on. Select inputs are identified as A, B, C and so on.

4.3.1 Operation

The operation of demultiplexer is explained using 1:4 DEMUX. The block diagram of 1:4 DEMUX with data input (D_{in}) from data source, select inputs (A and B) and the outputs (Z_0 to Z_3) of the DEMUX are shown in Fig. 4.8a. The truth table of the DEMUX is shown in Fig. 4.8b. An example of timing diagram for the DEMUX indicating the waveforms of select inputs, data input and outputs are shown in Fig. 4.8c.

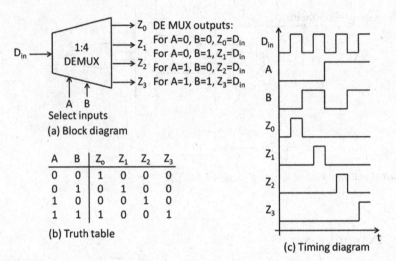

Fig. 4.8 Functioning of 1:4 DEMUX

4.3.1.1 Logic Diagram of 1:4 DEMUX

The logic diagram of 1:4 DEMUX is shown in Fig. 4.9. Four AND gates are shown in the logic diagram. The select inputs are A and B. Two inverters in the select input circuit generate the complements of A and B. The select inputs and D_{in} are connected to the AND gates. The outputs of the DEMUX are Z_0, Z_1, Z_2 and Z_3.

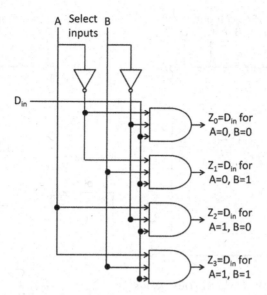

Fig. 4.9 Logic circuit diagram of 1:4 DEMUX

The select input circuit outputs four combinations of logic signals to the AND gates for routing the input data to the appropriate output line. The output, Z_0, is D_{in} for the select inputs, $AB = 00$; Z_1 is D_{in} for $AB = 01$; Z_2 is D_{in} for $AB = 10$ and Z_3 is D_{in} for $AB = 11$. The outputs and the select inputs are shown in Fig. 4.9.

4.3.2 Cascading Demultiplexers

Demultiplexer ICs are readily available in wide range of configurations with added features. The ICs have two or more enable (strobe) inputs for obtaining higher order demultiplexer by cascading demultiplexers. An example is presented.

4.3.2.1 Obtaining 1:32 DEMUX Using 1:16 DEMUXes

Two numbers of 1:16 DEMUX IC, 74154, are required for obtaining 1:32 DEMUX. The ICs are identified as U_1 and U_2. The IC has two active-Low enable (strobe) inputs, four select inputs (A, B, C and D) and sixteen active-Low outputs (Z_0 to Z_{15}). The input data signal, D_{in}, is connected to one of the strobe inputs of both the ICs. Enable input, EN_{in} is connected to the other strobe input of U_1 and it is connected through an inverter to the strobe input of U_2. The simplified connection diagram for obtaining 1:32 DEMUX is shown in Fig. 4.10. Although the select inputs and the data input, D_{in}, are shown separately for the ICs in the figure, they should be paralleled for operation.

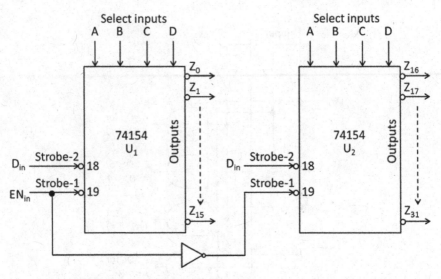

Fig. 4.10 1:32 DEMUX from 1:16 DEMUX ICs

When EN_{in} signal is 0, U_1 is enabled and D_{in} signal appears at the appropriate output terminal of U_1 as per the combination of select input signals. When EN_{in} signal is 1, U_2 is enabled and D_{in} signal appears at the appropriate output terminal of U_2 as per the combination of select input signals. As both data input and output pins have bubbles, the input and output data are inverted. D_{in} is realized at the outputs without inversion. Similarly, 1:16 DEMUX could be obtained by cascading two numbers of 1:8 DEMUX IC, 74138.

4.4 Decoders

A demultiplexer has one data input, n select inputs and less than or equal to 2^n outputs. A decoder is similar to demultiplexer except that it has no data input. Decoder with two select inputs and four (2^2) outputs is called 2-to-4 decoder. Decoder with three select inputs and eight (2^3) outputs is called 3-to-8 decoder and so on. One of the outputs of decoder goes High for each combination of select inputs.

4.4.1 Operation

The logic diagram of 2-to-4 decoder is shown in Fig. 4.11. Four AND gates are shown in the logic diagram. The select inputs are A and B. Two inverters in the select input circuit generate the complements of A and B. The select inputs and their complements are connected to the AND gates. The outputs of the decoder are Z_0, Z_1, Z_2 and Z_3.

The select input circuit outputs four combinations of logic signals to the AND gates. The output, Z_0, goes High for the select inputs, AB $= 00$; Z_1 goes High for AB $= 01$; Z_2 goes High for AB $= 10$ and Z_3 goes High for AB $= 11$. The outputs of the decoder and the combinations of the select inputs are shown in Fig. 4.11.

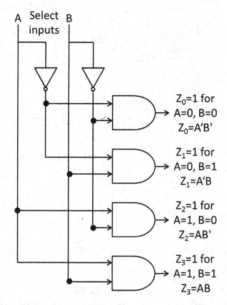

Fig. 4.11 Logic circuit diagram of 2-to-4 Decoder

4.4.2 Demultiplexers as Decoders

Demultiplexer ICs could be used as decoders by grounding its data input lines. IC 74154 is a Decoder/Demultiplexer. The method of using 74154 as 4-to-16 decoder is shown in Fig. 4.12. Similarly, IC 74138 could be used as 3-to-8 decoder.

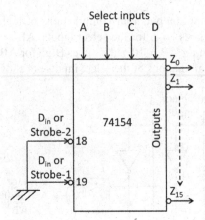

Fig. 4.12 Using DEMUX IC as decoders

4.4.3   Applications of Decoders

Standard ICs are available to support the applications of decoder. Four applications of decoder are presented.

(i)   Selecting memory systems and peripheral devices:
      Decoder is used for selecting memory systems and peripheral devices. The combinations of select inputs could be considered as address codes. The address codes could be linked to memory systems and peripheral devices such as printers and storage devices. When the address code of a device is applied to decoder, the logic High output of the decoder enables the device, linked to the address code.
(ii) Implementing logic functions.
(iii) BCD to Decimal Decoder/Driver.
(iv) BCD to Seven-Segment Decoder/Driver.

4.4.4   Implementing Logic Functions

Decoder is used for implementing the logic function of truth table. The number of select inputs of decoder should be equal to the number of variables of logic function. Assume that the truth table of 3-variable function, shown in Fig. 4.13a, should be implemented using 3-to-8 Decoder IC 74138. The logic function of the truth table is:

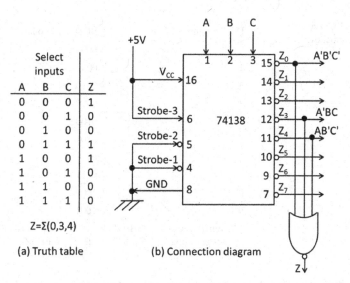

Fig. 4.13 Generating 3-variable logic function using 3-to-8 decoder

$$Z = \Sigma(0, 3, 4)$$

The connection diagram for implementing the logic function of the truth table using the IC, 74138, is shown in Fig. 4.13b. Strobe inputs are not needed for implementing logic function. Accordingly, active-High strobe input is connected to V_{CC} and active-Low strobe inputs are grounded. The outputs of the decoder at Z_0, Z_3 and Z_4 are the minterms m_0, m_3 and m_4 respectively. They are connected to NOR gate. The output (Z) of the NOR gate represents the logic function of the truth table.

4.4.5 BCD to Decimal Decoder/Driver

BCD is the acronym for Binary-Coded Decimal. BCD codes are 4-bit codes. They are also called BCD 8421 codes based on the weights of significance of the bits. The right most bit is called LSB (Least Significant Bit) and the left most bit is called MSB (Most Significant Bit). The weight of the LSB is 2^0 i.e. 1 and that of MSB is 2^3 i.e. 8. The weights of the in between bits are 2^1 (2) and 2^2 (4). Decimal numbers, 0 to 9, are represented using a 4-bit BCD code. The BCD codes for the decimal numbers are shown in Table 4.1.

Table 4.1 BCD codes for decimal numbers

Decimal number	BCD 8421 code
0	0000
1	0001
2	0010
3	0011
4	0100
5	0101
6	0110
7	0111
8	1000
9	1001

The remaining codes, 1010, 1011, 1100, 1101, 1110 and 1111 are not permitted in BCD 8421 code. They are invalid codes.

4.4.5.1 Standard ICs for BCD to Decimal Decoder/Driver

BCD to decimal decoder ICs (7442 and 7445) are available. They have four select inputs and ten inverted outputs (4-to-10 decoder) representing the decimal numbers 0 to 9. The decoder outputs drive ten LEDs, which are numbered 0 to 9 for displaying the decimal numbers. BCD input logic ensures that all outputs remain OFF for the invalid inputs, 1010, 1011, 1100, 1101, 1110 and 1111. The decoder IC, 7445, has high sink current capability for driving the LEDs directly. The datasheets of IC manufacturers could be referred for additional application information.

4.4.6   BCD to 7-Segment Decoder/Driver

Decimal numbers, 0 to 9, could be displayed by seven LEDs instead of using ten LEDs. The LEDs are arranged in the form of segments and the arrangement of the LEDs is called 7-segment display. Decoder with four select inputs and seven outputs (4-to-7 decoder) drives the seven LEDs. Seven-segment display and the operation of BCD to 7-Segement decoder are explained.

4.4.6.1 Seven-Segment Display

LEDs are used in wide range of applications such as indicators, displays and cell phone back lighting. The size and geometry of the p-n junction, and the fabrication process used, depend on the type of application. The area of LED chip for indicator applications is quite small. The fabrication of display LEDs in 7-segment displays is different from indicator LED chip. The p-n junction of display LEDs is in bar form obtaining very high line brightness from the edge of the junction; larger numeric LEDs are generally assembled from seven individual diodes [1].

Seven-segment displays are available in many sizes for various applications. The realization of the decimal numbers, 0 to 9, using 7-segment display is shown in Fig. 4.14. The seven LEDs are labelled from a to g. All the seven LEDs are used for displaying the decimal number, 8. All LEDs except the LED, g, are used for displaying the decimal number, 0; only the LEDs, b and c, are used for displaying the decimal number, 1; the LEDs, a, b, d, e and g are used for displaying the decimal number, 2, and so on as shown in the figure.

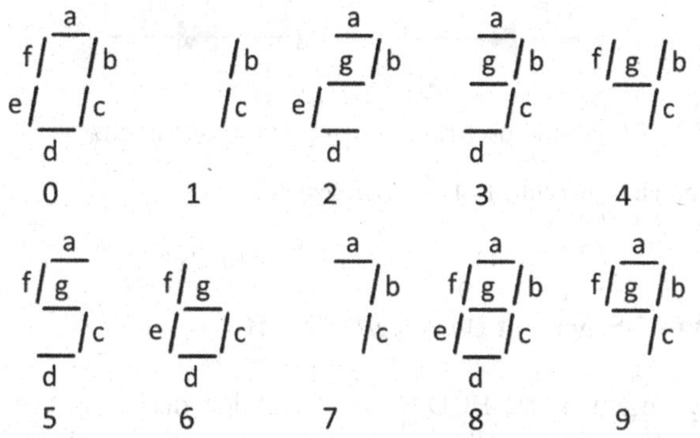

Fig. 4.14 Realization of decimal numbers using 7-sement display

4.4.6.2 Interconnection of LEDs in Seven-Segment Displays

Seven-segment display is a single device with pin-outs similar to ICs. Seven LEDs are electrically interconnected within the device. The interconnection of LEDs is available in two configurations, namely, common cathode and common anode. The cathodes of the seven LEDs are connected together and grounded in the common cathode configuration. The anodes of the seven LEDs are connected together to V_{CC} in the common anode configuration. Both the configurations are shown in Fig. 4.15. The LEDs are connected to the outputs of BCD to 7-segment decoder/driver ICs.

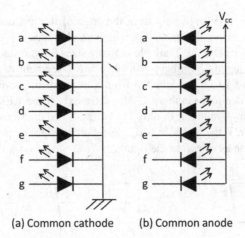

(a) Common cathode   (b) Common anode

Fig. 4.15 Types of interconnecting LEDs in 7-sement displays

4.4.6.3   BCD to 7-Segement Decoder/Driver ICs

There are many variants in the BCD to 7-segment decoder/driver ICs. In addition to decoder, the ICs have additional gate circuits to drive the required LED segments for displaying decimal numbers. The basic functional requirements remain same for all the ICs. The ICs have four select inputs (A, B, C and D) and seven outputs (Z_0 to Z_6). The common cathode configuration of 7-segement display is used with ICs having inverted outputs. The common anode configuration of the display is used with ICs having normal outputs. The connection diagram of the LEDs of seven-segment display with the ICs, 7447 and 7448 is shown in Fig. 4.16. Only the select inputs and outputs of the ICs are shown in the figure. The LED, a, is connected to the output, Z_0; LED, b, is connected to the output, Z_1; LED, c, connected to the output, Z_2 and so on. The ICs have additional features such as lamp intensity modulation, lamp test provision and leading/trailing zero suppression. Data sheets of IC manufacturers could be referred for complete details.

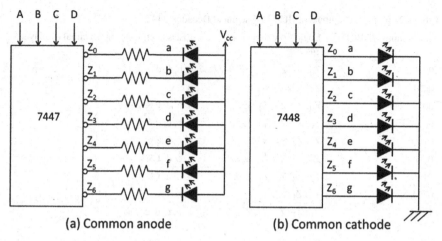

Fig. 4.16 BCD to 7-sement decoders/drivers

It is necessary to connect pull-up resistors in series with LEDs for protection. The external resistors are shown for the IC, 7447. The resistors are not shown for the IC, 7448, as the IC has built-in pull-ups eliminating the need for external resistors.

4.4.6.4 Operation of BCD to 7-Segement Decoder/Driver ICs

The operation of the BCD to 7-segment decoder/driver circuit in Fig. 4.16a is explained. It is applicable for the circuit in Fig. 4.16b. The IC, 7447, has inverted outputs. The LEDs of the common cathode configuration 7-segment display are connected to the outputs. Two or more outputs of BCD to 7-segment decoder/driver ICs go Low for each combination of select inputs. For the select input, $ABCD = 0000$, the outputs, Z_0 to Z_5, of the IC go Low and the LEDs, a, b, c, d, e & f, are energized displaying the decimal number, 0. For the select input, $ABCD = 0001$, the outputs, Z_1 and Z_2, of the IC go Low and the LEDs, b and c, are energized displaying the decimal number, 1 and so on. The combinations of select inputs (BCD 8421 codes), the logic Low outputs of the decoder, the LEDs that are energized and the decimal number displayed are shown in Table 4.2.

4.5 Encoders

A decoder has n inputs and less than or equal to 2^n outputs. An encoder has n outputs and less than or equal to 2^n inputs. The inputs of encoder are binary data representing decimal or octal numbers, alphabetical characters, symbols, etc. Encoders convert the selected i.e. active input data into BCD codes or another form of binary data appropriate to the input data. The operation of encoder for converting binary data representing decimal numbers into BCD codes is explained.

Table 4.2 Inputs and outputs of BCD to 7-segment Decoder, 7447

Select inputs (ABCD)	Logic Low outputs of 7447	LEDs energized	Decimal displayed
0000	Z_0, Z_1, Z_2, Z_3, Z_4 & Z_5	a, b, c, d, e & f	0
0001	Z_1 & Z_2	b & c	1
0010	Z_0, Z_1, Z_3, Z_4 & Z_6	a, b, d, e & g	2
0011	Z_0, Z_1, Z_2, Z_3 & Z_6	a, b, c, d & g	3
0100	Z_1, Z_2, Z_5 & Z_6	b, c, f & g	4
0101	Z_0, Z_2, Z_3, Z_5 & Z_6	a, c, d, f & g	5
0110	Z_2, Z_3, Z_4, Z_5 & Z_6	c, d, e, f & g	6
0111	Z_0, Z_1 & Z_2	a, b & c	7
1000	$Z_0, Z_1, Z_2, Z_3, Z_4, Z_5$ & Z_6	a, b, c, d, e, f & g	8
1001	Z_0, Z_1, Z_2 Z_5 & Z_6	a, b, c, f & g	9

4.5.1 Decimal-to-BCD Encoder

Key board is an input device to computers and it contains mechanical switches, numbered from 0 to 9. The switches are operated to input the decimal numbers. The decimal numbers are converted into BCD 8421 codes for computing applications by encoder. Four bits ($n = 4$) are required to represent the decimal numbers. Encoder with ten inputs i.e. 10-line to 4-line encoder is required. The simplified block diagram of encoder without logic gate circuit is shown in Fig. 4.17. The decimal number inputs from key board switches to the encoder are also shown in the figure. The outputs of the encoder are represented as ABCD. The MSB of the 4-bit output is A and the LSB is D.

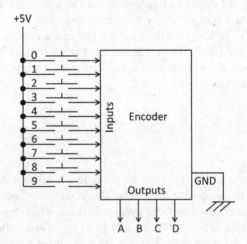

Fig. 4.17 Decimal-to-BCD Encoder

Active decimal input	Binary input to encoder	BCD output from encoder (ABCD)
0	1000000000	0000
1	0100000000	0001
2	0010000000	0010
3	0001000000	0011
4	0000100000	0100
5	0000010000	0101
6	0000001000	0110
7	0000000100	0111
8	0000000010	1000
9	0000000001	1001

Table 4.3 Inputs and outputs of Decimal-to-BCD encoder

4.5.1.1 Operation

Assume that the keyboard switch of the decimal number, 0, is operated. The input of the decoder is said to be active for the decimal number, 0. The binary input to the encoder is 1000000000. The output (ABCD) of the encoder is 0000. If the input of the decoder is made active for the decimal number, 1, the binary input to the encoder is 0100000000, the output (ABCD) of the encoder is 0001. The Decimal-to-BCD encoder converts active decimal inputs into BCD 8421 codes. The active decimal inputs, binary inputs and BCD outputs of the encoder for the decimal numbers are shown in Table 4.3.

4.5.1.2 Undefined Outputs

There are two sources of undefined outputs in the operation of Decimal-to-BCD encoder. Two or more keyboard switches could be operated providing more than one active inputs to the encoder. It might result in undefined outputs from the encoder.

The output of Decimal-to-BCD encoder is 0000 when none of the keyboard switches is operated i.e. when there are no active High inputs to the encoder. The same output is realized when the active High input of the decimal number, 0, is applied to the encoder. Realizing same output for two types of signal inputs result in undefined outputs. Both the sources of undefined outputs from encoders are eliminated in priority encoders.

4.5.2   Priority Encoders

Standard ICs are available for priority encoders. IC 74147 is Decimal-to-BCD priority encoder. IC 74148 is Octal-to-Binary priority encoder. The operation of the priority encoders is presented for understanding the methods of eliminating undefined outputs.

4.5.2.1   Decimal-to-BCD Priority Encoder

The Decimal-to-BCD (10-line to 4-line) priority encoder circuit using the IC, 74147 is shown in Fig. 4.18. The IC has bubbles both at inputs and outputs. The inputs representing decimal numbers are active Low signals and the outputs are inverted.

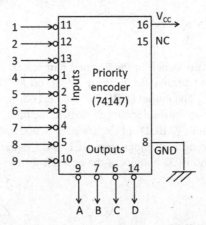

Fig. 4.18 Decimal-to-BCD priority encoder using IC 74147

If two or more active Low inputs of decimal numbers are applied to the priority encoder, the output of the encoder corresponds to the binary input representing the highest decimal number. For example, if the active Low inputs of the decimal numbers, 2, 5 and 9 are applied to the priority encoder, the output of the encoder is the inverted BCD code of the decimal number, 9, i.e. 0110 (LHHL). The inputs corresponding to the decimal numbers, 2 and 5, are treated as don't-care inputs. The datasheet of 74147 could be referred for the don't-care inputs and the outputs.

It could be observed in Fig. 4.18 that there are only nine input lines for the decimal numbers and there is no provision for the binary input representing the decimal number, 0, in the IC 74147. When none of the keyboard switches is operated, the priority encoder outputs the inverted BCD code of the decimal number, 0, i.e. 1111 (HHHH). The outputs of the encoder are defined for all the inputs of the encoder, eliminating undefined outputs.

4.5.2.2 Octal-to-Binary Priority Encoder

The method of eliminating undefined outputs in Octal-to-Binary priority encoder is different. The Octal-to-Binary priority encoder (IC 74148) has Group Select (GS) output control for eliminating undefined outputs. The IC has bubbles both at inputs and outputs. The inputs require active Low signals and the outputs are inverted. If GS is Low, it indicates that the output of the priority encoder corresponds to valid active Low input. If GS is High, it indicates that the output of the priority encoder is invalid.

When none of the keyboard switches is operated, all the inputs of the priority encoder are active High. The output of the priority encoder is 111 (HHH) and GS remains at High, indicating that the output is invalid. If the active Low input of the decimal number, 0, is applied to the priority encoder, the output of the priority encoder is 111 (HHH) and GS changes to Low, indicating that the output is valid.

If two or more active Low inputs of decimal numbers are applied to the priority encoder, the output of the encoder corresponds to the binary input representing the highest decimal number. GS remains at Low indicating that the output is valid.

4.6 Magnitude Comparators

Magnitude comparator compares two binary words having the same number of bits. The output of the comparator indicates whether one binary word is less than or equal to or greater than the other binary word. Examples of the applications of magnitude comparators are microprocessor based controllers and communication systems. The datasheets of manufacturers indicate that magnitude comparators are used for servo motor controls and process controllers.

4.6.1 Operation

Assume that the words, A and B, are to be compared. The words, A and B, are the inputs to magnitude comparator. The magnitude comparator has three outputs, namely, $(A < B)$, $(A = B)$ and $(A > B)$. The three outputs are denoted as L, E and G respectively. If $A < B$, the output, L, goes High. If $A = B$, the output, E, goes High. If $A > B$, the output, G, goes High. Theoretically, the number of bits in the two binary words could vary from one to n bits. Obtaining the logic functions of 1-bit and 2-bit magnitude comparators is explained. It could be extended for higher order magnitude comparators.

4.6.2   Logic Function for 1-Bit Comparator

Consider two 1-bit words having A_0 and B_0 as the bits of the words. If A_0 is less than B_0, the output L_0 goes High. If A_0 is equal to B_0, the output E_0 goes High. If A_0 is greater than B_0, the output G_0 goes High. The functional block diagram and truth table for the inputs and outputs of 1-bit magnitude comparator is shown in Fig. 4.19. The simplified logic functions in SOP form for L_0, E_0 and G_0, are:

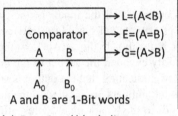

Inputs		Outputs		
A_0	B_0	L_0 (A<B)	E_0 (A=B)	G_0 (A>B)
0	0	0	1	0
0	1	1	0	0
1	0	0	0	1
1	1	0	1	0

(a) Functional block diagram   (b) Truth table

Fig. 4.19 1-Bit magnitude comparator

$$F(A < B) = L_0 = A_0'B_0$$
$$F(A = B) = E_0 = A_0'B_0' + A_0B_0$$
$$F(A > B) = G_0 = A_0B_0'$$

4.6.3   Logic Function for 2-Bit Comparator

Let the two 2-bit words be A and B. The bits of the words are A_1A_0 and B_1B_0. The MSBs are A_1 and B_1. The LSBs are A_0 and B_0. The generic truth table comparing MSBs and LSBs is shown in Fig. 4.20. The truth table contains two don't-care inputs (X) as the outputs of comparator are decided by MSBs alone. For example, if A_1 is greater than B_1, then A is greater than B irrespective of the logic values of the LSBs. Similarly, if A_1 is less than B_1, then A is less than B irrespective of the logic values of the LSBs. Hence, X is assigned for the comparing input, (A_0, B_0) for the two cases.

Comparing inputs		Outputs		
A_1, B_1	A_0, B_0	A < B	A = B	A > B
$A_1 > B_1$	X	0	0	1
$A_1 = B_1$	$A_0 > B_0$	0	0	1
$A_1 = B_1$	$A_0 = B_0$	0	1	0
$A_1 = B_1$	$A_0 < B_0$	1	0	0
$A_1 < B_1$	X	1	0	0

Fig. 4.20 Generic truth table of 2-bit magnitude comparator

4.6.3.1 Obtaining Logic Functions

There are five sets of comparing inputs in Fig. 4.20. The comparing inputs are used for obtaining the logic functions for the three outputs of the 2-bit comparator.

F (A < B):
The output, (A < B), goes High for two sets of the comparing inputs. The comparing inputs are:

(i) When $A_1 < B_1$
(ii) When $A_1 = B_1$ and $A_0 < B_0$

$$F(A < B) = F(A_1 < B_1) + F(A_1 = B_1) \cdot F(A_0 < B_0)$$
$$F(A < B) = L_1 + E_1 \cdot L_0$$

F (A = B):
The output, (A = B), goes High for one set of the comparing inputs and the comparing input is:

(i) When $A_1 = B_1$ and $A_0 = B_0$

$$F(A = B) = F(A_1 = B_1) \cdot F(A_0 = B_0)$$
$$F(A = B) = E_1 \cdot E_0$$

F (A > B):
The output, (A > B), goes High for two sets of the comparing inputs and the comparing inputs are:

(i) When $A_1 > B_1$
(ii) When $A_1 = B_1$ and $A_0 > B_0$

$$F(A > B) = F(A_1 > B_1) + F(A_1 = B_1) \cdot F(A_0 > B_0)$$
$$F(A > B) = G_1 + E_1 \cdot G_0$$

4.6.4   Cascading Magnitude Comparators

Magnitude comparators are cascaded to obtain higher order comparators in digital processing applications. For example, 4-Bit comparator could be cascaded to obtain 8, 12, 16 and other higher order comparators. Standard ICs (Ex.: 7485 and CD4063) are available for comparing various sizes of words. Schematic diagram for cascading comparators and truth table are available in the datasheets of ICs. Realizing 2-Bit comparator by cascading two 1-Bit comparators is presented.

4.6.4.1   Cascading 1-Bit Comparators

Let the 2-Bit words to be compared be A and B. The bits of the words are A_1A_0 and B_1B_0. The functional block diagram cascading two 1-bit comparators are shown in Fig. 4.21. The comparators are marked as LSBC (Less Significant Bits Comparator) and MSBC (More Significant Bits Comparator). The bits, A_0 and B_0, are the inputs (A_L and B_L) for LSBC. The bits, A_1 and B_1, are the inputs (A_M and B_M) for the MSBC.

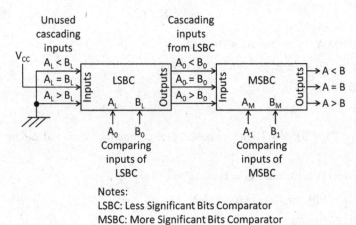

Fig. 4.21 Cascading 1-Bit magnitude comparators

Comparators have provision at their inputs for cascading comparators. The inputs to the ICs are called cascading inputs. The generic identification of cascading inputs are A_L and B_L. Cascading inputs are not needed for LSBC. The cascading inputs of LSBC are connected to V_{CC} and ground as shown in Fig. 4.21. The outputs of LSBC are the cascading inputs to MSBC. The cascading inputs to MSBC are A_0 and B_0. The MSBC outputs the final results of comparing the 2-bit words, A and B. One of the three outputs of MSBC goes High as per the result of comparison.

4.6.4.2   Truth Table of 2-Bit Comparator

The truth table of 2-bit comparator is shown in Fig. 4.22. It contains comparing inputs, cascading inputs and final outputs. Don't-care inputs (X) are shown in the table as appropriate for the cascading inputs. For example, if A_1 is greater than B_1, A is greater than B irrespective of the logic values for the LSBs. Hence, X is assigned for the cascading inputs. If A_1 is equal to B_1, the output is decided considering each of the cascading inputs.

Comparing inputs	Cascading inputs			Outputs		
A_1, B_1	$A_0 < B_0$	$A_0 = B_0$	$A_0 > B_0$	$A < B$	$A = B$	$A > B$
$A_1 > B_1$	X	X	X	0	0	1
$A_1 = B_1$	0	0	1	0	0	1
$A_1 = B_1$	0	1	0	0	1	0
$A_1 = B_1$	1	0	0	1	0	0
$A_1 < B_1$	X	X	X	1	0	0

Fig. 4.22 Generic truth table of 2-Bit comparator

4.6.4.3   Cascading 4-Bit Comparators

LSBC and MSBC are explained for cascading higher order magnitude comparators. 4-Bit comparators are cascaded for comparing two 8-bit words, A and B. The 4-bit data inputs to LSBC are denoted as A_L and B_L and the inputs are the less significant bits, $A_3A_2A_1A_0$ and $B_3B_2B_1B_0$ of the words. The 4-bit data inputs to MSBC are denoted as A_M and B_M and the inputs are the more significant bits, $A_7A_6A_5A_4$ and $B_7B_6B_5B_4$. The truth table of cascaded 4-bit comparators is similar to the truth table of 2-bit comparator in Fig. 4.22. The words of LSBC and MSBC are compared in preparing the truth table of 8-bit comparator. The datasheets of 4-bit comparators could be referred for additional information.

Reference

1. Turner LW (ed) (2013) Electronics engineer's reference book, 4th edn. Butterworths-Heinemann

Chapter 5
Number Systems and Binary Codes

Abstract Decimal numbers are converted into binary numbers for digital processing applications. In addition to binary number system, octal and hexadecimal number systems are also used for digital processing. Apart from numeric data, digital signal processing requires analog data. Binary codes are used to represent analog data. Number systems and binary codes are explained. When binary codes are transmitted, bit errors might be observed at receiving end. Bit error detection methods, namely, single parity check, two dimensional parity check, check sum and cyclic redundancy, are presented.

5.1 Types of Number Systems

Humans use decimal number system in daily life for counting and other mathematical calculations. The number system has the digits, 0 to 9. Decimal number system is not suitable for computers and other microprocessor related applications. Computers use binary number system, having the digits, 0 and 1. Decimal numbers are converted into binary numbers for digital processing applications.

In addition to binary number system, octal and hexadecimal number systems are also used for digital processing considering the complexity of processing needs. Hexadecimal is studied because it is a useful means to represent large sets of binary values using a manageable number of symbols [1]. The number systems, their conversions and binary coding are presented.

5.2 Decimal Number System

The characteristics of decimal number system are well known. The next number is always obtained by incrementing the previous number by one. The radix or the base of the decimal number system is 10 as the system contains ten digits from 0 to 9. The radix representation (powers of 10) of a decimal number indicates the positional

© Springer Nature Switzerland AG 2020
D. Natarajan, *Fundamentals of Digital Electronics*,
Lecture Notes in Electrical Engineering 623,
https://doi.org/10.1007/978-3-030-36196-9_5

weight of the digit in the decimal number. As an example, consider the four digit decimal number, 4689.567. The positional weight of decimal number is used in the radix representation.

$$\text{Decimal number} \quad 4 \quad 6 \quad 8 \quad 9 \quad 5 \quad 6 \quad 7$$
$$\text{Positional weight} \; 10^3 \; 10^2 \; 10^1 \; 10^0 \; 10^{-1} \; 10^{-2} \; 10^{-3}$$
$$4689.576 = 4 \times 10^3 + 6 \times 10^2 + 8 \times 10^1 + 9 \times 10^0 + 5 \times 10^{-1}$$
$$+ 6 \times 10^{-2} + 7 \times 10^{-3}$$

It is obvious that the weight of the digit, 4, is higher than the digit, 9, in the whole number. Similarly, the weight of the digit, 5 is higher than the digit, 7 in the fractional part of the number. In general, the weight of more significant digit is higher than the weight of less significant digit.

The decimal point which separates the whole number and fraction is called radix point. The characteristics of the decimal number system are applicable to binary, octal and hexadecimal number systems except that the radix (base) values vary between the number systems.

5.3   Binary Number System

The binary number system contains the digits, 0 and 1. The radix of binary number system is 2. The term, bit, is used instead of digit in binary system. Generally, binary numbers have 3 or more bits. The characteristics of binary number system are explained for 4-bit number. The first number is 0000. The next number is 0001 and it is obtained by adding 1 to 0000. The next number is 0010 and the addition process could be continued until the largest 4-bit number, 1111, is obtained. The 4-bit numbers from 0000 to 1111 are shown in Table 5.1. The right most bit of binary number is the least significant bit (LSB). The left most bit of the number is the most significant bit (MSB).

5.3.1   Decimal-Binary Conversion

Decimal number is converted into binary number by dividing the whole number of the decimal number repetitively by the radix of binary system. Repetitive division by 2 is performed for the conversion. Repetitive multiplication by 2 is performed on the fractional part of the decimal number for conversion. Decimal to binary conversion is illustrated for the decimal number, 11.6875. The converted binary number for the decimal number is 1011.1011.

Table 5.1 4-bit binary numbers and their equivalent decimal numbers		
4-bit binary number	**Equivalent decimal number**
0000	0
0001	1
0010	2
0011	3
0100	4
0101	5
0110	6
0111	7
1000	8
1001	9
1010	10
1011	11
1100	12
1101	13
1110	14
1111	15

5.3.1.1 Converting Whole Number of Decimal into Binary

The whole number of the decimal number, 11 is divided by 2, obtaining quotient and remainder. The remainder is the LSB of the binary number of the decimal number. Repetitive division by 2 is performed on the quotients obtained after every division and the operation is continued until the quotient of 0 is obtained. The remainder obtained for the quotient of 0 is the MSB of the binary number. The binary number for the decimal number is read from the MSB to LSB. The application of the method is illustrated for converting the decimal number, 11 into binary number in Fig. 5.1a. The binary equivalent number for the decimal number, 11 is also indicated in the figure.

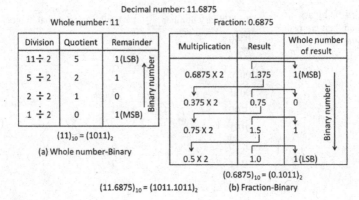

Fig. 5.1 Decimal-Binary conversion

5.3.1.2 Fractional Decimal-Binary Conversion

The fractional part of the decimal number, 0.6875 is multiplied by 2 and the result is 1.375. The whole number of the result is 1 and it is the MSB of the equivalent fractional binary number. The multiplication process is continued for the fractional part of the decimal number, 1.375. The multiplication of 0.375 by 2 gives 0.75. The whole number is 0 and it is the first lower bit from MSB. The multiplication of 0.75 by 2 gives 1.5. The whole number is 1 and it is the second lower bit from MSB. The multiplication of 0.5 by 2 gives 1.0. As the fractional part of the decimal number, 1.0, is 0, the multiplication process is stopped. The whole number is 1 and it is the third lower bit of MSB i.e. LSB. The conversion process is shown in Fig. 5.1b. The binary equivalent number for the fractional part of the decimal number is also indicated in the figure.

The fractional part of the decimal number might not become 0 even after repetitive multiplication by 2. In such cases, the multiplication process is stopped after obtaining the required number of binary digits for the fractional part.

5.3.2 Binary-Decimal Conversion

The radix representation of 4-bit binary numbers indicates the weight of the bits in the number. The positional weight of decimal numbers is applicable for binary numbers also. Binary to decimal conversion is illustrated separately for the whole number and the fractional part of the binary number, 1001.1011. The converted decimal number for the binary number is 9.6875.

5.3.2.1 Converting Whole Number of Binary into Decimal

The whole number part of the binary number is 1001. LSB has the least weight and MSB has the largest weight. The sum of the radix representation of binary number gives the equivalent decimal number. The radix representation of the 4-bit number, 1001 is:

$$(1001)_2 = 1 \times 2^3 + 0 \times 2^2 + 0 \times 2^1 + 1 \times 2^0 = (9)_{10}$$

5.3.2.2 Converting Binary Fraction to Decimal

The fractional part of the binary number is 0.1011. The sum of the radix representation of binary fraction gives the equivalent decimal fraction. The radix representation of the 4-bit binary fraction considering the weight of the bits is:

$$(0.1011)_2 = 1 \times 2^{-1} + 0 \times 2^{-2} + 1 \times 2^{-3} + 1 \times 2^{-4} = (0.6875)_{10}$$

5.3.3 Binary Coded Decimal

Decimal numbers could also be represented in Binary Coded Decimal (BCD) format. The most popular code is the BCD-8421 code. BCD codes are used in many applications although higher complexity of hardware is need for mathematical operations. With the rapid advances in very large scale integration (VLSI) technology, semi- and fully parallel hardware decimal multiplication units with BCD encoding are expected to evolve soon [2]. Four bits are used for the representation of BCD 8421 code.

The BCD-8421 codes of the decimal digits, 0 to 9, are obtained using the same procedure in Sect. 5.3.1. The codes are tabulated in Table 5.2 and they are used for representing decimal numbers larger than 9 in BCD-8421 code format. The codes, 1010 to 1111 are considered invalid in BCD-8421 code.

5.3.3.1 Decimal to BCD-8421 Code

A decimal number is represented in BCD-8421 code format by converting each digit of the decimal number into 4-bit binary number. Alternatively, Table 5.2 could be used. An example for converting the decimal number, 25.6, into BCD-8421 code is shown in Fig. 5.2a. The converted BCD code for the decimal number is 100101.011 after omitting leading and trailing zeroes.

Table 5.2 BCD-8421 codes
for the decimal numbers 0 to 9

Decimal number	BCD-8421 code
0	0000
1	0001
2	0010
3	0011
4	0100
5	0101
6	0110
7	0111
8	1000
9	1001

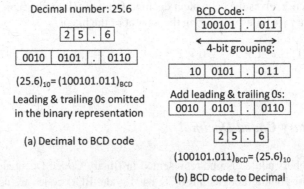

Fig. 5.2 Example: Decimal-BCD-Decimal conversions

5.3.3.2 BCD-8421 Code to Decimal

An example for obtaining the equivalent decimal number for the BCD-8421 code, 100101.011, is shown in Fig. 5.2b. The reverse operations are performed for obtaining the equivalent decimal number of a BCD-8421 code format. The binary code is re-written in groups of 4 bits starting from the radix point (decimal point) as shown by the direction of the arrows in the figure. The grouping is applicable both for the whole number and the fractional part of the binary code. Leading and trailing 0s are added to the 4-bit groups as applicable. Each 4-bit group is converted into its equivalent decimal digit using Table 5.2. The converted decimal number for the BCD code, 100101.011 is 25.6.

5.3.3.3 Understanding Conversion and Coding

The bits of binary numbers after converting decimal numbers have positional values. The bits of BCD code format for decimal numbers do not have positional values. The BCD format for a decimal number is actually coding and not conversion. For example, the equivalent decimal number of the BCD code, 100101.011, is 25.6. Applying positional values to the BCD code, 100101.011, the equivalent decimal number is 37.375. Positional values are not applicable for binary code formats. More information regarding binary codes is available in Sect. 5.6.

5.3.4 Excess-3 Code

Excess-3 code is another BCD format for representing decimal numbers. Excess-3 code is obtained by adding 0011 (decimal equivalent, 3) to each BCD-8421 code. Usually, an Excess-3 code is used when one desires to perform arithmetic operations by the method of compliments [3]. Arithmetic logic circuits are simpler with Excess-3 codes compared to BCD-8421codes. BCD-8421 codes and their equivalent Excess-3 codes are indicated in Table 5.3.

Table 5.3 Excess-3 and BCD codes

Decimal number	BCD-8421 code	Excess-3 code (BCD+0011)
0	0000	0011
1	0001	0100
2	0010	0101
3	0011	0110
4	0100	0111
5	0101	1000
6	0110	1001
7	0111	1010
8	1000	1011
9	1001	1100

5.4   Octal Number System

Octal number system has eight digits from 0 to 7. The next octal number after 7
is obtained by adding 1 to 7. The sum, $(7 + 1)$ is equal to 10. The octal numbers
continue from 10 to 17 and the next octal number after 17 is obtained by adding 1
to 17. The sum, $(17 + 1)$ is equal to 20. The octal numbers continue from 20 to 27
and the next octal number after 27 is 30. The octal number series continue similarly.
The radix of octal numbers is 8 and it is used for converting octal numbers to other
number systems.

Converting between 2^n bases (e.g., 2, 4, 8, 16) takes advantage of the direct
mapping that each of these bases back to binary; Base 8 numbers take exactly 3
binary bits to represent all 8 symbols; Base 16 numbers take exactly 4 binary bits
to represent all 16 symbols [1]. As n is equal to 3 in octal number system, each
basic octal number is represented by 3 binary bits. The basic octal numbers and their
equivalent decimal and binary numbers are shown in Table 5.4. The table is used for
converting larger octal numbers into other number systems and vice versa.

5.4.1   Octal-Binary Conversion

Assume that the octal number, 153.74 should be converted into binary number. Each
octal digit is converted into binary number using the Table 5.4. The conversion is
shown in Fig. 5.3a. The leading and trailing zeroes are omitted after conversion. The
converted binary code for the octal number is 1100011.1111. The positional values
for the bits of the binary code after conversion are not applicable.

Table 5.4 Octal and their equivalent numbers

Basic octal number	Equivalent decimal number	Equivalent 3-bit binary number
0	0	000
1	1	001
2	2	010
3	3	011
4	4	100
5	5	101
6	6	110
7	7	111

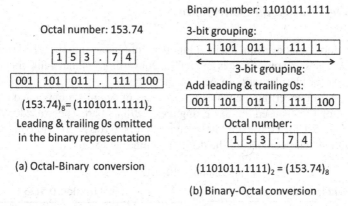

Fig. 5.3 Example: Octal-Binary-Octal conversion

5.4.2 Binary-Octal Conversion

An example for converting the binary code, 1101011.1111 into octal number is shown in Fig. 5.3b. The reverse operations are performed in converting a binary number into octal number. The binary code is re-written in groups of 3 bits starting from the radix point (decimal point) as shown by the direction of the arrows in the figure. The grouping is applicable both for the whole number and the fractional part of the binary code. Leading and trailing 0s are added to the 3-bit groups as applicable. Each 3-bit group is converted into its equivalent octal digit using Table 5.4. The converted octal number for the binary code, 1100011.1111 is 153.74.

5.4.3 Octal-Decimal Conversion

The radix of the octal number system is 8. The sum of the radix representation of octal number considering the positional weight of the digits in the number is used for obtaining the equivalent decimal number. Consider the octal number, 152.64. The equivalent decimal number for the octal number is:

$$152.64 = 1 \times 8^2 + 5 \times 8^1 + 2 \times 8^0 + 6 \times 8^{-1} + 4 \times 8^{-2} = 106.8125$$
$$(152.64)_8 = (106.8125)_{10}$$

5.4.4   Decimal-Octal Conversion

The procedure for converting decimal numbers into binary numbers is described in Sect. 5.3.1. The same procedure is applied for converting decimal numbers into octal numbers except that the radix of octal number system is used. Repetitive division by 8 is used for converting the whole number part of decimal number and repetitive multiplication by 8 is used for converting the fractional part. The procedure is illustrated in Fig. 5.4 for converting the decimal number, 106.8125 into octal number. The converted octal number for the decimal number is 152.64.

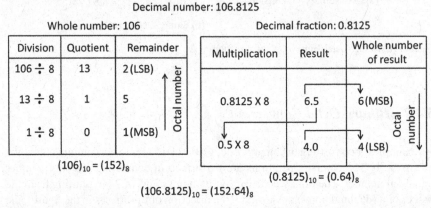

Fig. 5.4   Example: Decimal-Octal conversion

The fractional part of the decimal number might not become 0 even after repetitive multiplication by 8. In such cases, the multiplication process is stopped after the required number of fractional octal digits.

5.5   Hexadecimal Number System

Hexadecimal number system has a total of sixteen digits and the digits are 0 to 9 and A to F. The next hexadecimal number after F is obtained by adding 1 to F. The sum, (F + 1) is equal to 10. The hexadecimal numbers continue from 10 to 19 and from 1A to 1F. The next hexadecimal number after 1F is obtained by adding 1 to 1F. The sum, (1F + 1) is equal to 20. The hexadecimal numbers continue from 20 to 29 and from 2A to 2F. The next hexadecimal number after 2F is 30. The hexadecimal number series continue similarly. The radix of hexadecimal numbers is 16 and it is used for converting hexadecimal numbers to other number systems.

As the radix of hexadecimal numbers is the fourth power of 2, four bits are used to represent each basic hexadecimal number. The basic hexadecimal numbers and their equivalent decimal and binary numbers are shown in Table 5.5. The table is used for converting larger hexadecimal numbers into other number systems and vice versa.

Table 5.5 Hexadecimal and their equivalent numbers	Basic hexadecimal number	Equivalent decimal number	Equivalent 4-bit binary number
	0	0	0000
	1	1	0001
	2	2	0010
	3	3	0011
	4	4	0100
	5	5	0101
	6	6	0110
	7	7	0111
	8	8	1000
	9	9	1001
	A	10	1010
	B	11	1011
	C	12	1100
	D	13	1101
	E	14	1110
	F	15	1111

5.5.1 Hexadecimal-Binary Conversion

Each hexadecimal digit is converted into binary number using Table 5.5. The leading and trailing zeroes are omitted after conversion. An example for converting the hexadecimal number, 37F.2E, into binary code is shown in Fig. 5.5a. The converted binary code for the hexadecimal number is 1101111111.0010111 after omitting leading and trailing zeroes. Positional values for the bits of the binary code after conversion are not applicable.

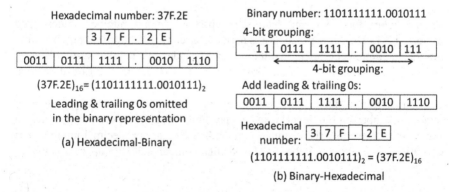

(a) Hexadecimal-Binary

(b) Binary-Hexadecimal

Fig. 5.5 Example: Hexadecimal-Binary-Hexadecimal conversions

5.5.2   Binary-Hexadecimal Conversion

An example for converting the binary code, 1101111111.0010111 into hexadecimal number is shown in Fig. 5.5b. The reverse operations are performed in converting a binary number into hexadecimal number. The binary code is re-written in groups of 4 bits starting from the radix point (decimal point) as shown by the direction of the arrows in the figure. The grouping is applicable both for the whole number and the fractional part of the binary code. Leading and trailing 0s are added to the 4-bit groups as applicable. Each 4-bit group is converted into its equivalent hexadecimal digit using Table 5.5. The converted hexadecimal number for the binary code is 37F.2E.

5.5.3   Hexadecimal-Decimal Conversion

The radix of the hexadecimal number system is 16. The sum of the radix representation of octal number considering the positional weight of the digits in the number is used for obtaining the equivalent decimal number. Consider the octal number, 37F.2E. Table 5.5 is used for obtaining the decimal values of the alphabets in hexadecimal numbers. The equivalent decimal number for the hexadecimal number is:

$$37F.2E = 3 \times 16^2 + 7 \times 16^1 + 15 \times 16^0 + 2 \times 16^{-1}$$
$$+ 14 \times 16^{-2} = 895.18 \text{ approx.}$$
$$(37F.2E)_{16} = (895.18)_{10}$$

5.5.4   Decimal-Hexadecimal Conversion

The procedure for converting decimal numbers into binary numbers is described in Sect. 5.3.1. The same procedure is applied for converting decimal numbers into hexadecimal numbers except that the radix of hexadecimal number system is used. Repetitive division by 16 is used for converting the whole number part of decimal number and repetitive multiplication by 16 is used for converting the fractional part. The multiplication process for converting decimal fraction is stopped after the required number of fractional hexadecimal digits. The procedure is illustrated in Fig. 5.6 for converting the decimal number, 895.18 into hexadecimal number. The converted hexadecimal number for the decimal number is 37F.2E approximately.

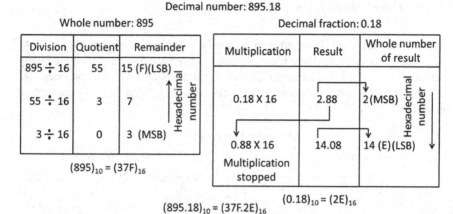

Fig. 5.6 Example: Decimal-Hexadecimal conversion

5.5.5 Hexadecimal-Octal Conversion

Two steps are involved in converting hexadecimal number into octal number. First the hexadecimal number is converted into binary number as per Sect. 5.5.1 and then the binary number is converted into octal number as per Sect. 5.4.2. The two steps are illustrated in Fig. 5.7 for converting the hexadecimal number, 37F.2E into octal number. The converted octal number for the hexadecimal number is 1577.134.

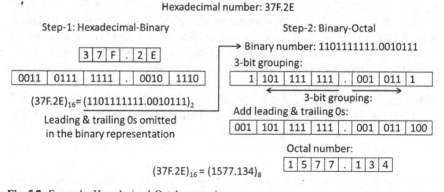

Fig. 5.7 Example: Hexadecimal-Octal conversion

5.5.6 Octal-Hexadecimal Conversion

Two steps are involved in converting octal number into hexadecimal number. First the octal number is converted into binary number as per Sect. 5.4.1 and then the binary number is converted into hexadecimal number as per Sect. 5.5.2. The two steps are illustrated in Fig. 5.8 for converting the octal number, 1577.134 into hexadecimal number.

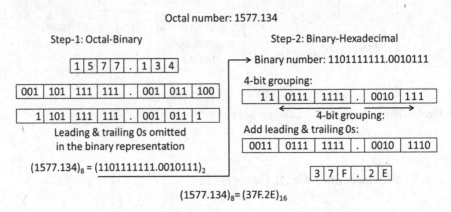

Fig. 5.8 Example: Octal-Hexadecimal conversion

5.6 Binary Codes

Numeric data is one of the inputs for digital signal processing. Apart from numeric data, digital signal processing requires analog data. Analog data is the core input for digital signal processing. Characters and analog variables are examples of analog data. Analog input variables could be DC voltages (or currents) or modulated AC waveforms using a variety of modulation techniques or in some combination, with a spatial configuration of related variables to represent shaft angles [4]. Analog data is converted into binary code formats using standard Analog-to-Digital Converter ICs.

Many types of binary code formats are available for representing analog data. Binary code formats of data contain 0s and 1s in combinational form. Each code format is unique in its form. Appropriate code format is used considering the requirements of conversion from analog to digital, processing digital signals and re-conversion from digital to analog signal. The following types of binary code formats are presented:

(i) Unipolar binary codes

Unipolar binary codes are used for representing positively varying analog voltages after conversion [4]. The popular codes in data conversion are straight binary code and Gray code. The codes are presented.

(ii) Bipolar binary codes

Bipolar binary codes are used for representing analog signals (Ex. Sinusoidal signal) that vary positively and negatively after conversion. Either offset binary, twos complement, ones complement, or sign magnitude codes will accomplish this, but offset binary and two's complement are by far the most popular [4]. Offset binary and two's complement codes are presented.

5.6.1 Unipolar Straight Binary

The application of straight binary code is presented for DC voltage that varies from 0 V to some specified maximum voltage. The maximum voltage is called full scale voltage. The straight binary coding for the DC voltage:

- The zero voltage is assigned to the code, 0000
- The full scale voltage is assigned to the code, 1111
- The intermediate voltages are assigned to the other codes from 0001 to 1110.

Standard analog-to-digital converter ICs such as ADS7842 are available and the ICs output straight binary codes after conversion. The straight binary codes are same as the equivalent binary numbers for the decimal numbers, 0 to 15. The codes count from 0000 to 1111.

5.6.2 Unipolar Gray Code

Gray code is used for representing positively varying analog signal in digital format after conversion. As an example, the analog-to-digital converter for converting the mechanical displacements of shaft rotation into digital signals outputs Gray code and the code groups are converted into 4-bit binary or BCD codes for further processing [3]. 4-bit Gray code is shown in Table 5.6. If required, Gray code could be converted into straight binary code using standard gates.

Table 5.6 Gray codes

4-bit Gray code
0000
0001
0011
0010
0110
0111
0101
0100
1100
1101
1111
1110
1010
1011
1001
1000

5.6.3   Bipolar Offset Binary

Offset binary code is used for representing the converted output of positively and negatively varying voltage. 4-bit Offset binary codes are shown in Table 5.7. Assume that the voltage of analog signal varies from $-V_{FS}$ (negative full scale voltage) to $+V_{FS}$ (positive full scale voltage). The offset binary coding scheme for the analog signal is:

- The zero voltage is assigned to the code, 1000
- The negative voltage levels are assigned to the codes from 0111 down to 0000 $(-V_{FS})$
- The positive voltage levels are assigned to the codes from 1001 up to 1111 $(+V_{FS})$.

Standard analog-to-digital converter ICs such as ADS7800 and ADS5231 are available and the ICs output offset binary codes after conversion.

5.6.4   Bipolar Binary Two's Complement

Binary Two's complement coding is the type of coding used by most microprocessor or math processor based systems for mathematical algorithms, and is also the coding scheme which the digital audio industry has decided to use as its standard [5]. 4-bit Binary two's complement codes are shown in Table 5.6. Assume that the voltage of

Table 5.7 Offset binary codes

4-bit offset binary code
0000
0001
0010
0011
0100
0101
0110
0111
1000
1001
1010
1011
1100
1101
1110
1111

analog signal varies from $-V_{FS}$ (negative full scale voltage) to $+V_{FS}$ (positive full scale voltage). The offset binary coding scheme for the analog signal is:

- The zero voltage is assigned to the code, 0000
- The negative voltage levels are assigned to the codes from 1111 down to 1000 $(-V_{FS})$
- The positive voltage levels are assigned to the codes from 0001 up to 0111 $(+V_{FS})$.

Standard analog-to-digital converter ICs such as ADS5231 are available and the ICs output binary two's complement codes after conversion (Table 5.8).

5.6.4.1 Conversion Between Offset Binary and Two's Complement

Offset binary code could be converted into Two's complement code by inverting the MSB of the offset binary code and vice versa. The output data of the standard IC, ADS5231 is available either in offset binary or in two's complement format.

5.7 Alphanumeric Codes

Apart from representing numbers (decimal, octal and hexadecimal) and analog signals in binary format, characters are also represented in binary format. Alphanumeric codes are used for encoding characters. Characters are alphabets, punctuation marks,

Table 5.8 Two's
complement codes

4-bit two's complement code
1000
1001
1010
1011
1100
1101
1110
1111
0000
0001
0010
0011
0100
0101
0110
0111

digits (0 to 9), strings (Ex. NUL), symbols (Ex. $, &), etc. The alphanumeric code schemes contain digits and they are different from decimal numbers. The digits are used in programming languages.

Encoding schemes such as Hollerith code on punched cards for characters existed for computers in earlier days. Three alphanumeric code schemes exist for encoding characters. The code schemes are ASCII (American Standards Code for Information Interchange), EBCDIC (Extended Binary Coded Decimal Interchange Code) and Unicode. Character representation on computing systems became less dependent on hardware; instead, software designers used the existing encoding schemes to accommodate the needs of an increasingly global community of computer users [6].

5.7.1   ASCII and EBCDIC Schemes

In 1963, American Standards Association (ASA) announced the American Standard Code for Information Interchange (ASCII) scheme [6]. ASCII uses seven bits for representing characters. ASCII code was extended to eight bits for accommodating additional characters. The most widely used Single-Byte Character Sets (SBCS) encoding today, after ASCII, is ISO-8859-1 and it is an 8-bit superset of ASCII and provides most of the characters necessary for Western Europe [6].

IBM developed the 8-bit EBCDIC code scheme. It was used on the successful IBM System/360 mainframe computer series, which hit the market in April 1964 [6].

ASCII and EBCDIC code pages could be referred for the binary codes of characters in the schemes.

5.7.2 Unicode Standard

As the population of world is a multilingual society, the characters of many linguistic societies need to be represented in binary format. Unicode standard is used for representing the characters. Xerox Corporation's Star Workstation, which had a multilingual word processor called ViewPoint, and IBM Corporation's 5550 office computers could process multiple Asian languages, in addition to multiple languages that use the Latin script led to the Unicode movement that created a multilingual character set [7]. Unicode is the universal character encoding standard used for the representation of text for computer processing and the standard has been adopted by industry leaders Apple, HP, IBM, Microsoft, Oracle, SAP, Sun, Sybase, Unisys [8].

5.8 Bit Error Detection

Various digital sub-systems like servers and storage devices are interconnected using transmission lines. Examples of transmission lines are coaxial and twisted pair cables. Appropriate transmitters are used at one end for sending binary coded message data along transmission line. The transmitted data is received at the other end of the line. The received data is reconstructed for use. Sending and receiving binary codes also occur between in-circuit digital blocks.

Message data is transmitted in the order of hundreds of Mbps (Megabits per second). Factors such as external transients, attenuation in transmission lines and cross talk between lines alter the status of the bits of transmitted message data. Logic zero in the message data might be received as logic one and vice versa. One or more bits could be altered randomly. The unwanted logic state changes in message data are called bit errors. Random bit errors in the transmission path affect the quality of received data. Preventive actions such as the use of shielded cables and repeaters are taken to avoid random bit errors in received data. However, it is necessary to check for random bit errors in the received data.

5.8.1 Types of Random Bit Errors

Random bit errors are classified as single-bit error and burst error. One of the bits of transmitted data might be found altered when the data is received and the error is called single-bit error. If two or more number of bits are found altered in received data, the error is called burst error. Examples of the errors in received data are shown in Fig. 5.9. Altered bits are shown with fills.

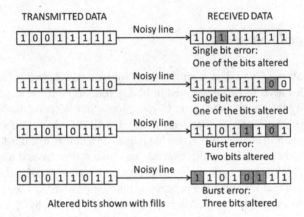

Fig. 5.9 Random bit errors

5.8.2 Error Detection Methods

Many error detection methods are available. Algorithms are available for implementing error detection methods. Appropriate method is selected considering the application needs of data communication and the probability of the occurrence of bit errors. The general approach for detecting bit errors in received message is common to all the methods.

Additional (redundant) bits are generated at the transmitting end and they are sent with message data for detecting errors. The additional bits are used at the receiving end for detecting bit errors in message data. If received message data contains bit errors, recall information is sent to transmitters for sending the data again. Three error detection methods, namely, parity checking, checksum and cyclic redundancy checking, are presented.

5.8.2.1 Parity Checking

Parity checking is generally used for bit oriented protocols such as high level data link control (HDLC) and local area networks (LANs). There are two methods of parity checking for error detection. Single bit parity checking and two dimensional parity checking methods are used for bit error detection. Both the methods are presented.

5.8.3 Single Bit Parity Check

An extra bit either 0 or 1 is added to message data at the transmitting end for detecting bit errors that occur in the transmission path. The extra bit (redundant bit) is called parity bit. For example, if the message data contains four bits, the transmitted data contains five bits including parity bit; if the message data contains eight bits, the transmitted data contains nine bits including parity bit and so on. The parity bit is used at the receiving end for detecting errors in message data.

There are two ways of adding parity bit to message data. Parity bit could be added to make the number of 1s in transmitted data (message data + parity bit) odd or even. If message bit contains two 1s, transmit data with odd parity is obtained by adding 1 to the message bit. If message bit contains three 1s, transmit data with odd parity is obtained by adding 0 to the message bit.

Similarly, transmit data with even parity is obtained by adding 1 or 0 to the message bit. Examples of transmitted data with odd parity and even parity are shown in Fig. 5.10. The added parity bits are identified in the figure. Parity generators are used to transmit data with odd or even parity. Parity checkers are used at the receiving end for detecting errors in message data.

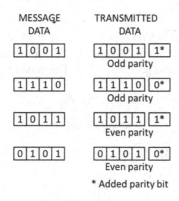

Fig. 5.10 Transmitted data with parity bits

5.8.3.1 Parity Generators

Parity bits are generated by using XOR and XNOR gate circuits. The gate circuit is called parity generator. As an example, the gate circuit for generating odd parity bit to 3-bit message data, ABC (101) is shown in Fig. 5.11a. The transmission data is 1011.

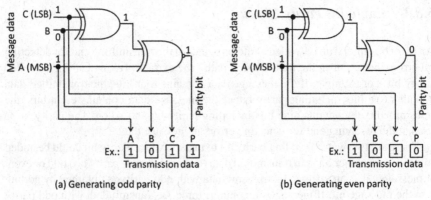

Fig. 5.11   Gate circuits for generating parity bits

Similarly, the gate circuit for generating even parity bit to 3-bit message data, ABC (101) is shown in Fig. 5.11b. The transmission data is 1010. Standard ICs (Ex. 74180 and 74280) for parity generators are available.

5.8.3.2   Parity Checkers

Parity checker for odd parity 3-bit receiver system is shown in Fig. 5.12. XOR and XNOR gates are used in the circuit. Assume that the transmitted data with odd parity bit is ABCP (1011). The message data is 101. The added odd parity bit is 1.

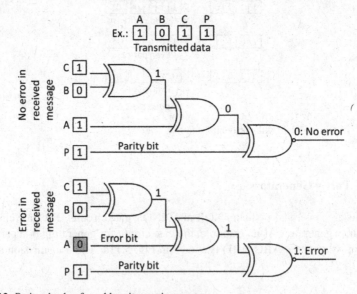

Fig. 5.12   Parity checker for odd parity receiver system

If received data is 1011, the output of the parity checker is 0, indicating no bit errors in the received data. If the received data is 0011, the output of the parity checker is 1, indicating bit error in the received data. Both the types of received data and the outputs of parity checkers are shown in the figure.

The XNOR gate at the output of the circuit could be replaced by XOR gate and the circuit becomes parity checker for even parity 3-bit receiver system. Standard ICs (Ex. 74180 and 74280) for parity checkers are available.

5.8.3.3 Limitations

Single bit parity checker method can also detect burst errors with odd number of changes in the bits of message data. Referring to Fig. 5.12, if the received data is 0101 for the transmitted message 1011, the error is detected by the parity checker. If the received data is 1101, the error is not detected. Two dimensional parity checking, checksum and cyclic redundancy check (CRC) techniques are capable of detecting most of the burst errors with varying levels of confidence. The overview of the functioning of the error detection techniques is presented.

5.8.4 Two Dimensional Parity Check

Two dimensional parity check method generally uses even parity system for a group of message data. The message data is arranged in the form an array. Even parity bits are identified for the rows and columns of the message array. The application of two dimensional parity check method is illustrated for identifying even parity bits on 7-bit message data.

Five sets of 7-bit message data (ABCDEFG) are arranged in the form of array and the array is shown in Fig. 5.13. Even parity bits are identified for the five sets of 7-bit message data and the row parity bits are shown in the figure.

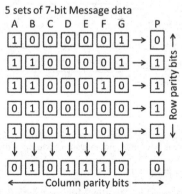

Transmitted data: Five sets of message data
with parity bits and column parity bits (8 bits)

Fig. 5.13 Two dimensional even parity check

The next step is to identify even parity bit for columns. To begin with, even parity bit is identified for MSBs (most significant bits) of the five sets of 7-bit message data and the column parity bit is recorded as shown in the figure. Identifying column parity bits is continued until the LSBs (least significant bits) of message data are covered. Finally, the five row parity bits are treated as a column. The column parity bit of the five row parity bits is also identified, covering the five sets of 7-bit message data. The one set of eight column parity bits are shown in the figure.

Data with parity bits is transmitted. The transmitted data contains the five sets of message data with row parity bits and the one set of eight column parity bits. The parity bits are used for detecting errors in received message. Errors affecting four bits may not be detected by two dimensional parity check method [9].

5.8.5   Checksum

Two types of binary operations are performed on the sets of message data in the checksum error detection method. Binary addition is performed on the sets of message data using 1's complement arithmetic to obtain the sum of the sets of message data. Checksum bits are obtained by complementing the sum. The checksum bits are the redundant bits for error detection and they are sent with message data. The checksum bits are used to detect errors in received message data.

The application of checksum method is illustrated using three sets of 8-bit message data in Fig. 5.14. Message data-1 and Message data-2 are added and the sum of the codes of the data is obtained. The carry bit resulting from the addition is added to the sum in 1's complement (inverse) arithmetic. The intermediate sum after the addition of the carry bit is shown in the figure.

```
Message data-1    1 0 0 0 0 0 1 0
Message data-2  + 1 1 0 0 0 0 1 1
                  ─────────────────
                  0 1 0 0 0 1 0 1
Carry from data-1+data-2        + 1
                  ─────────────────
Intermediate sum  0 1 0 0 0 1 1 0
Message data-3  + 1 0 0 1 1 0 0 1
                  ─────────────────
Final sum         1 1 0 1 1 1 1 1

Checksum =        0 0 1 0 0 0 0 0
```
Notes:
Checksum = 1's complement of final sum
Carry is 0 for Intermediate sum + Message data-3

Fig. 5.14 Checksum for error detection

The intermediate sum is added to the Message data-3 for obtaining final sum. Carry is 0 for the addition of intermediate sum and Message data-3. Checksum is the 1's complement of the final sum. Checksum is transmitted with message data. It is used for detecting errors in received message. Checksum method detects close to 95% of the errors for multiple-bit burst errors [10].

5.8.6 Cyclic Redundancy Check

Cyclic redundancy check for bit error detection is popularly called CRC. One of the most popular error-checking schemes is CRC and CRC-32 (a 32-bit version) will detect about 99.99999998% of all burst errors longer than 32 bits [10]. CRC error detection method assumes an appropriate polynomial and the message bits are assigned as the coefficients of the polynomial. Binary division is used for obtaining CRC bits. The message data and the CRC bits are transmitted. The CRC bits are used for detecting errors in received message.

The algorithm of CRC is integrated with the design of digital blocks. The application report of Texas Instruments presents software algorithms and provides examples of implementing CRC [11]. CRC is explained in tutorial form in Ref. [12].

References

1. LaMeres BJ (2017) Introduction to logic circuits & logic design with VHDL, Springer, 2017
2. Jaberipur G, Kaivani A (2007) Binary-coded decimal digit multipliers. IET Comput Dig Tech 1(4):377
3. Oluwade D (2012) A comparative analysis and application of the compression properties of two 7-bit subsets of unicode. J Emerg Trends Comput Inf Sci 3(4)
4. Walt K (ed) (2004) Analog-digital conversion, analog devices, Mar 2004
5. Application report (2015) Application report SBAA042A, Texas instruments, May 2015
6. How unicode relates to prior standards such as ASCII and EBCDIC, IBM Knowledge Centre
7. Steven JS (2004) A brief history of character codes in North America, Europe and East Asia. Sakamura Laboratory, University Museum, University of Tokyo
8. Vining B (2007) What's with these ASCII, EBCDIC, Unicode CCSIDs, IBM Corporation
9. Forouzan AB (2007) Data communication & network. McGraw Hill
10. Fitzgerald J, Dennis A, Durcikova A (2012) Business data communications and networking. Wiley & Sons Inc.
11. Geremia P (1999) Cyclic redundancy check computation: an implementation using the TMS320C54X. Application report SPRA530, Texas Instruments, Apr 1999
12. Matloff N (2001) Cyclic redundancy checking. University of California, Davis

Chapter 6
Arithmetic Operations and Circuits

Abstract Arithmetic operations are part of digital signal processing. The arithmetic operations, binary addition and subtraction, are illustrated with examples. The combinational logic circuits for performing addition and subtraction are presented. Binary multiplication and division are illustrated with examples.

6.1 Binary Arithmetic Operations

The Central Processing Unit (CPU) of a computer consists of Arithmetic Logic Unit (ALU) and Control Unit. ALU carries out arithmetic operations and logical functions as per the instructions of Control Unit. The arithmetic operations are binary addition and subtraction. The logical functions are OR, AND, NOT, XOR, NOR and other required functions. Appropriate combinational logic circuits for performing the arithmetic operations and logical functions are part of ALU. The arithmetic operations, binary addition and subtraction, are illustrated with examples. The combinational logic circuits for performing addition and subtraction are also presented.

Binary multiplication and division are also arithmetic operations. Recursive algorithms are generally used for the operations. ALU performs the operations with additional sequential logic circuits. Standard ICs (Ex. CD4089B) are available for performing binary multiplication and division. CD4089B performs the operations in conjunction with an up/down counter and control logic. Only the recursive algorithms for performing binary multiplication and division are illustrated with examples. 4-bit binary numbers are used for illustrating binary arithmetic operations.

6.2 Binary Addition

Addition of decimal numbers is performed if the numbers are positive i.e. unsigned. Consider the addition of two decimal numbers, 58 and 24. The addition of the two decimal numbers, 58 and 24, is shown in Fig. 6.1. The digits at 10^0 positions are

© Springer Nature Switzerland AG 2020
D. Natarajan, *Fundamentals of Digital Electronics*,
Lecture Notes in Electrical Engineering 623,
https://doi.org/10.1007/978-3-030-36196-9_6

added first and then the digits at 10^1 positions are added. The digits at 10^0 positions are 8 and 4. The result of adding 8 and 4 is 12. The digit, 2, of the result is entered at 10^0 position and it is termed as sum digit. The addition generates the carry digit, 1. The carry digit is entered at 10^1 position. The carry digit, 1, and the digits, 5 and 2, are then added. The result of the addition is 8 with no carry digit and it is entered at 10^1 position. The addition of unsigned binary numbers is similar to the addition of decimal numbers.

$$
\begin{array}{l}
\text{Carry digit} \longrightarrow 1 \\
\qquad\qquad\quad 5\ \ 8 \\
\qquad\qquad\quad 2\ \ 4 \\
\hline
\text{Sum digits} \longrightarrow 8\ \ 2
\end{array}
$$

Fig. 6.1 Addition of decimal numbers

6.2.1   Addition of 4-Bit Binary Numbers

The terms, sum bit and carry bit, are used in binary addition. The addition procedure is applicable for the converted binary numbers of decimal numbers, BCD codes and the binary representation of hexadecimal and octal numbers. The rules of binary addition are derived from Boolean algebra.

6.2.1.1   Rules for Binary Addition

The rules for binary addition are specified for adding two 1-bit binary numbers, A_0 and B_0. The rules are:

$0 + 0 = 0$ (The sum bit is 0 with no carry bit i.e. the carry bit is 0)
$0 + 1 = 1$ (The sum bit is 1 and the carry bit is 0)
$1 + 0 = 1$ (The sum bit is 1 and the carry bit is 0)
$1 + 1 = 10$ (The sum bit is 0 at 2^0 position and the carry bit is 1 at 2^1 position)

Three bits might have to be added in the addition of higher order binary numbers. The derived rule for adding three bits is:

$1 + (1 + 1) = 1 + 10 = 11$ (The sum and carry bits are 1 at 2^0 and 2^1 positions).

6.2.1.2   Illustrations

The number of bits of unsigned binary numbers decides the range of decimal numbers that could be added. In general, the range of decimal numbers is given by 0 to

$(2^n - 1)$, where n is the number of bits in the binary numbers. The range of decimal numbers that could be added by 4-bit binary numbers is 0 to 15.

The addition of 4-bit numbers is illustrated with three examples in Fig. 6.2. Addition begins with LSBs and continued until the addition of MSBs is completed. The sum bits and the carry bits for the examples are shown in the figure. Carry bit is relevant for 2^1 position and higher. If the carry bit is not indicated, it is 0. The equivalent decimal numbers are also shown against the binary numbers in the figure. If binary number contains fractional part, addition begins with the LSBs of the fractional part until the addition of MSBs of whole number is completed.

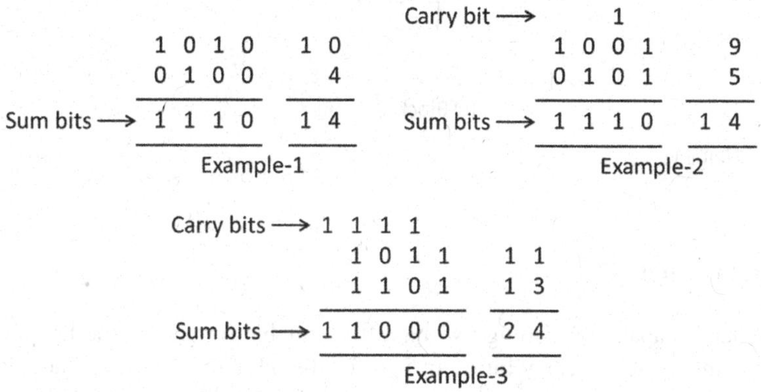

```
                                    Carry bit ⟶        1
            1 0 1 0    1 0                    1 0 0 1         9
            0 1 0 0       4                   0 1 0 1         5
Sum bits ⟶  1 1 1 0      1 4     Sum bits ⟶  1 1 1 0       1 4
                Example-1                         Example-2


            Carry bits ⟶ 1 1 1 1
                         1 0 1 1      1 1
                         1 1 0 1      1 3
            Sum bits ⟶ 1 1 0 0 0      2 4
                         Example-3
```

Fig. 6.2 Addition of binary numbers

6.2.1.3 Logic Circuits for Addition

Combinational logic circuits are used for performing binary addition. The operation of Half adder, Full adder, Parallel adder and Fast adder is presented. Standard ICs are also available for binary addition.

6.2.2 Half Adder

The logic circuit that adds two bits (A_0 and B_0) is called half adder. The rules of adding two 1-bit binary numbers are shown in the form of truth table in Fig. 6.3a. The simplified logic function for the truth table is obtained using K-map. The combinational logic circuit diagram for implementing the logic function is shown in Fig. 6.3b. The block diagram of half adder is shown in Fig. 6.3c.

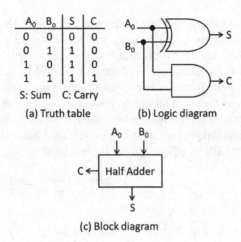

(a) Truth table   (b) Logic diagram

(c) Block diagram

Fig. 6.3 Half adder

6.2.2.1 Limitations

Half adder is capable of adding two bits. It is suitable for adding the LSBs of the binary numbers as the carry bit input is 0 for the addition of LSBs. The addition of LSBs might generate carry bit (1). The carry bit should be added with the next higher two bits of LSBs. Half adder is not suitable for adding three bits. Another logic circuit that is capable of adding three bits is required.

6.2.3   Full Adder

The logic circuit that could add three bits is called full adder. Let A_i and B_i represent the bits of two binary numbers. Assume that C_i is the carry bit to be added with A_i and B_i. The truth table of full adder for adding the three bits is shown in Fig. 6.4a. The outputs of the truth table are the sum bit, S_i and the carry bit, $C_{(i+1)}$. The carry bit, $C_{(i+1)}$, is generated by the addition of A_i, B_i and C_i. The generated carry bit could be 0 or 1. The simplified logic function for the truth table is obtained using K-map. The combinational logic circuit diagram for implementing the logic function is shown in Fig. 6.4b. The logic circuit consists of two half adders (HA) and one OR gate. The block diagram of full adder (FA) is shown in Fig. 6.4c.

A_i	B_i	C_i	S_i	$C_{(i+1)}$
0	0	0	0	0
0	0	1	1	0
0	1	0	1	0
0	1	1	0	1
1	0	0	1	0
1	0	1	0	1
1	1	0	0	1
1	1	1	1	1

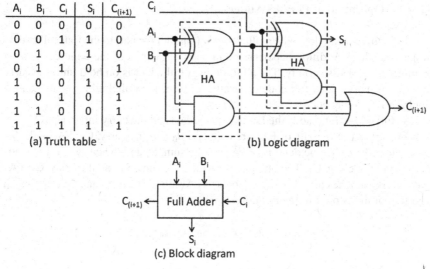

(a) Truth table   (b) Logic diagram

(c) Block diagram

Fig. 6.4 Full adder

6.2.4 Parallel Adder

Parallel adder is used for adding two higher order binary numbers. It is constructed by cascading full adders. The number of full adders for cascading is equal to the number of bits in the binary numbers that are added by parallel adder. The connection diagram for adding two 4-bit binary numbers using parallel adder is shown in Fig. 6.5. FA_0 to FA_3 are the four full adders.

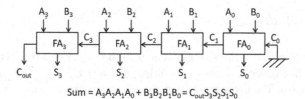

Sum $= A_3 A_2 A_1 A_0 + B_3 B_2 B_1 B_0 = C_{out} S_3 S_2 S_1 S_0$

Fig. 6.5 4-bit binary Parallel adder

Assume that the binary numbers, $A_3 A_2 A_1 A_0$ and $B_3 B_2 B_1 B_0$, are to be added by the parallel adder. The input pairs of bits of the binary numbers, the carry bits (inputs and outputs) and the output sum bits are shown in the figure. The full adder, FA_0, for adding the LSBs (A_0 and B_0) of the binary numbers could be replaced by half adder as C_0 is grounded.

6.2.4.1   Operation of Parallel Adder

The 4-bit numbers are loaded simultaneously (parallel format) to the four full adders of parallel adder. Addition is performed in pairs of the bits of the numbers starting from LSBs to MSBs serially. The addition by a full adder is on hold until the carry input from its previous full adder is available. Hence, parallel adder is also called ripple carry parallel adder.

The full adder, FA_0, adds the LSBs (A_0 and B_0) and the carry input, C_0 ($C_0 = 0$) to begin with. The outputs of FA_0 are the sum, S_0 and the carry, C_1. FA_1 adds A_1, B_1 and the carry, C_1. The outputs of FA_1 are the sum, S_1 and the carry, C_2. FA_2 adds A_2, B_2 and the carry, C_2. The outputs of FA_2 are the sum, S_2 and the carry, C_3. FA_3 adds A_3, B_3 and the carry, C_3. The outputs of FA_3 are the sum, S_1 and the carry, C_{out}. The sum of the two 4-bit binary numbers is:

$$A_3A_2A_1A_0 + B_3B_2B_1B_0 = C_{out}S_3S_2S_1S_0$$

6.2.4.2   Limitations

Parallel adder has limitation in the speed of adding two binary numbers as full adders perform addition serially after obtaining carry from their previous full adders. Finite gate delay exists for a full adder to obtain the carry from its previous full adder. The cumulative delays for obtaining the sum of two 16-bit or 32-bit binary numbers would be considerable and the delay may not be acceptable. Fast carry parallel adder (or simply fast adder) provides the solution for obtaining the sum of binary numbers with minimum gate delays.

6.2.5   Fast Adder

Consider 4-bit fast adder circuit for adding two 4-bit binary numbers. The construction of 4-bit fast adder is similar to 4-bit parallel adder. Full adders are cascaded in fast adder also. The carry inputs that would be generated during the addition of the 4-bit numbers are made available in advance (ahead) to the full adders before loading the numbers to the full adders. Additional logic circuit is used to generate the carry inputs in advance. The additional logic circuit is called look-ahead carry logic circuit. The binary numbers are then loaded to the full adders. Addition is performed on the binary numbers with minimum gate delay as the full adders need not wait for carry input from previous full adders. Fast adder is also called look-ahead carry adder. The derivation of the logic function for implementing the look-ahead carry logic circuit for the addition of two 4-bit numbers is presented. The derivations are based on carry generation and carry propagation.

6.2.5.1 Carry Generation and Propagation

Assume that A_i, B_i and C_i are the inputs of a full adder. A_i and B_i are the bits of binary numbers. C_i is the carry input to the full adder. The outputs of the full adder are the sum bit, S_i, and carry bit, $C_{(i+1)}$. The output carry, $C_{(i+1)}$, is 1, for two sets of inputs to the full adder. The two sets of inputs are:

(i) When both the inputs (A_i and B_i) of full adder are 1, the carry output, $C_{(i+1)}$, of the full adder is 1. In other words, the full adder generates carry, $C_{(i+1)}$, 1. The general logical expression for the generated carry is:

$$C_{(i+1)} = A_i B_i$$

(ii) When A_i or B_i is 1 and C_i is 1, the carry output, $C_{(i+1)}$, of the full adder is 1. In other words, the full adder propagates the input carry, C_i, as its output carry, $C_{(i+1)}$. The general logical expression for the propagated carry is:

$$C_{(i+1)} = (A_i + B_i)C_i$$

Let G_i is equal to $A_i B_i$ and P_i is equal to $(A_i + B_i)$. They are used for obtaining the logic functions of look-ahead carry circuit.

6.2.5.2 Logic Functions for Look-Ahead Carry Circuit

The parallel adder circuit in Fig. 6.5 is used for obtaining the logic functions for the look-ahead carry circuit of fast adder. The carry output of each full adder is obtained using G_i and Pi.

Output carry, C_1, of the Full adder FA_0:

$G_0 = A_0 B_0$
$P_0 = A_0 + B_0$
$C_1 = G_0 + P_0 C_0$

Output carry, C_2, of the Full adder FA_1:

$G_1 = A_1 B_1$
$P_1 = A_1 + B_1$
$C_2 = G_1 + P_1 C_1 = G_1 + P_1 (G_0 + P_0 C_0) = G_1 + P_1 G_0 + P_1 P_0 C_0$

Output carry, C_3, of the Full adder FA_2:

$G_2 = A_2 B_2$
$P_2 = A_2 + B_2$
$C_3 = G_2 + P_2 C_2 = G_2 + P_2 (G_1 + P_1 G_0 + P_1 P_0 C_0) = G_2 + P_2 G_1 + P_2 P_1 G_0 + P_2 P_1 P_0 C_0$

Output carry, C_{out}, of the Full adder FA_3:

$$G_3 = A_3B_3$$
$$P_3 = A_3 + B_3$$
$$C_{out} = G_3 + P_3C_3 = G_3 + P_3 (G_2 + P_2G_1 + P_2P_1G_0 + P_2P_1P_0C_0)$$
$$C_{out} = G_3 + P_3G_2 + P_3P_2G_1 + P_3P_2P_1G_0 + P_3P_2P_1P_0C_0$$

It could be observed that the expressions for the output carry bits (C_1, C_2, C_3 and C_{out}) of full adders are reduced as the functions of the pair of the input bits (A_i and B_i) of binary numbers and C_0. The functions are implemented by the look-ahead carry logic circuit. The logic circuit provides the input and output carry bits (C_1, C_2, C_3 and C_{out}) in advance before loading the 4-bit binary numbers to full adders. When the 4-bit binary numbers are loaded to the fast adder, the addition of the numbers is performed by the full adders with minimum gate delays.

6.2.5.3   Standard ICs for Fast Adders

All standard ICs (Ex. 74HC283 and MC14008B) for performing addition are fast adders. The look-ahead carry logic circuit is integrated with the adder circuit of the ICs. The data sheets of the ICs could be referred for look-ahead carry logic circuit. The functional block diagram of fast adder with the logic circuit is also available in the data sheets of ICs. The datasheets of the ICs specify gate delays and provide examples of computing gate delays for the addition of binary numbers using fast adders. They could be referred for additional information.

6.2.6 Cascading Fast Adders

Lower order fast order ICs are cascaded for adding two higher order binary numbers. For example, 4-bit fast adder ICs could be cascaded for adding 8-bit or 16-bit binary numbers. The application of 4-bit fast adder, 74HC283, for adding two 8-bit binary numbers ($A_7A_6A_5A_4A_3A_2A_1A_0$ and $B_7B_6B_5B_4B_3B_2B_1B_0$) is shown in Fig. 6.6. C_{in} is 0 for the IC adding the LSBs of the binary numbers and hence, it is grounded. V_{CC} and the ground connections of the ICs are not shown in the figure.

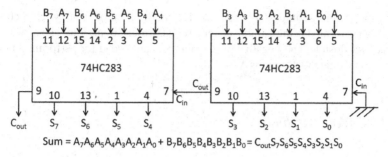

$$\text{Sum} = A_7A_6A_5A_4A_3A_2A_1A_0 + B_7B_6B_5B_4B_3B_2B_1B_0 = C_{out}S_7S_6S_5S_4S_3S_2S_1S_0$$

Fig. 6.6 Adding 8-bit numbers using 4-bit fast adder ICs

6.3 Binary Subtraction

Subtraction is performed on two decimal numbers when one of the two numbers is negative i.e. signed. Consider the two decimal numbers, 1053 and −478. The number, 1053, is called minuend and the number, −478 is called subtrahend. The digits of subtrahend are subtracted from the digits of minuend. The result after subtraction is called difference. Subtraction begins with the digits at the lowest position and it is continued up to the digits at highest positions. If the digit of minuend is less than the digit of subtrahend at any position, subtraction is performed after borrowing from higher positions to compute the difference.

6.3.1 Rules for Binary Subtraction

The decimal subtraction procedure is applicable for the converted binary numbers of decimal numbers, BCD codes and the binary representation of hexadecimal and octal numbers. The rules of binary subtraction are similar to the borrow method for subtracting decimal numbers. The rules for subtracting two 1-bit binary numbers, A_0 and B_0, are:

$0 - 0 = 0$ (The difference bit is 0 with no need to borrow)
$1 - 0 = 1$ (The difference bit is 1 with no need to borrow)
$1 - 1 = 0$ (The difference bit is 0 with no need to borrow)
$0 - 1 = 10 - 1 = 1$ (The difference bit is 1 after borrowing 1 from higher position).

6.3.1.1 Method for Subtraction

Appropriate logic circuits could be implemented as per the rules of the borrow method for performing binary subtraction. However, borrow method is not used as the method requires complex logic circuits for implementation. Binary subtraction is generally performed using parallel adder circuits, which simplify logic hardware for subtraction. Positive and negative binary numbers are coded for performing binary subtraction using adder circuits.

6.3.2   Coding Methods

Coding methods are applicable for the converted binary numbers of decimal numbers, BCD çodes and for the binary representation of hexadecimal and octal numbers. Signed magnitude, one's complement and two's complement methods are used for coding binary numbers. Signed magnitude and one's complement coding methods are briefly introduced as they have limitations in their applications. Two's complement coding method is widely used for coding binary numbers. Coded binary numbers are used for all arithmetic operations. Only addition and subtraction operations on the binary numbers of decimal numbers are explained using two's complement codes. Examples are also provided for the operations.

6.3.3   Signed Magnitude

A sign bit is added at the MSB position of binary number in the signed magnitude coding method. The MSB (sign bit) of binary number indicates whether the equivalent decimal number of binary number is positive or negative. Binary number with 0 at MSB position represents positive decimal number. Binary number with 1 at MSB position represents negative decimal number. For example, $+5_{10}$ is represented as 0 followed by the binary equivalent of 5 i.e. 0101. Similarly, -5_{10} is represented as 1 followed by the binary equivalent of 5 i.e. 1101. The magnitude of equivalent decimal number is obtained by discarding the MSB of binary number.

The range of decimal numbers is that could be represented by signed magnitude coded binary number is given by $-[2^{(n-1)} - 1]$ to $+[2^{(n-1)} - 1]$, where n is the number of bits in the binary number. The range of decimal numbers that could be represented by 4-bit (n = 4) signed magnitude code is -7 to $+7$. The representation of the decimal numbers in 4-bit signed magnitude code is shown in Table 6.1.

Table 6.1 Signed magnitude and one's complement codes for decimal numbers

Decimal number	Signed magnitude	One's complement
+7	0111	0111
+6	0110	0110
+5	0101	0101
+4	0100	0100
+3	0011	0011
+2	0010	0010
+1	0001	0001
+0	0000	0000
−0	1000	1111
−1	1001	1110
−2	1010	1101
−3	1011	1100
−4	1100	1011
−5	1101	1010
−6	1110	1001
−7	1111	1000

6.3.3.1   Limitations

Signed magnitude BCD is popular in bipolar digital voltmeters, but has the problem of two allowable codes for zero; it is therefore unpopular for most applications involving ADCs or DACs [1]. The decimal, 0, has two allowable codes in the signed magnitude coding method. +0 is represented by 0000 and −0 is represented by 1000. Additional logic circuits are needed for identifying sign bits for performing binary arithmetic operations.

6.3.4   One's Complement

Positive decimal number is represented by adding the bit, 0, as the MSB of its equivalent binary number in one's complement coding method. The remaining bits of the binary number represent the magnitude of the decimal number. For example, $+5_{10}$ is represented as 0 followed by the binary equivalent of 5 i.e. 0101. Negative decimal number is represented by the complement of the one's complement code of its positive decimal number. For example, the one's complement code of -5_{10} is $0'1'0'1'$ i.e. 1010.

The range of decimal numbers is that could be represented by one's complement coded binary number is given by $-[2^{(n-1)} - 1]$ to $+[2^{(n-1)} - 1]$, where n is the number of bits in the binary number. The range of decimal numbers that could be represented by 4-bit (n = 4) one's complement code is -7 to $+7$. The representation of the decimal numbers in 4-bit one's complement code is shown in Table 6.1.

6.3.4.1 Converting One's Complement Code to Decimal Number

It could be observed in Table 6.1 that the MSB of the one's complement codes of positive decimal numbers is 0 and that of the negative numbers is 1. The equivalent decimal number of one's complement codes is obtained by appropriate decoding operations. Consider the one's complement code, 0101. The MSB, 0, indicates that the equivalent decimal number is positive. The MSB bit is discarded and the equivalent decimal number of 101 is $+5_{10}$.

For the one's complement code, 1010, the MSB, 1, indicates that the equivalent decimal number is negative. The MSB bit is discarded. The remaining bits of the binary number are first complemented. Complement of 010 is 101. The equivalent magnitude of decimal number of 101 is 5_{10}. The equivalent decimal number of 1010 is -5_{10}.

6.3.4.2 Limitations

The decimal, 0, has two allowable codes in one's complement method also. $+0$ is represented by 0000 and -0 is represented by 1111. The two allowable codes need to be considered in data conversion. If the numbers of one's complement code are incremented beyond the largest value in the set, they roll over and start counting at the lowest number and the roll over is a useful feature for computer systems [2]. Adding 1 to 0111 gives 1000 illustrating the roll over feature.

6.3.5 Two's Complement

Positive decimal number is represented by adding the bit, 0, as the MSB of its equivalent binary number in two's complement coding method. The remaining bits of the binary number represent the magnitude of the decimal number. For example, $+5_{10}$ is represented as 0 followed by the binary equivalent of 5 i.e. 0101. Negative decimal number is represented by adding 1 to the complement of the two's complement code of its positive decimal number. For example, the two's complement code of -5_{10} is $(0'1'0'1' + 1)$ i.e. 1011.

The range of decimal numbers is that could be represented in two's complement coded system is given by $-[2^{(n-1)}]$ to $+[2^{(n-1)} - 1]$, where n is the number of bits in the binary codes. The range of decimal numbers that could be represented by 4-bit (n = 4) two's complement codes is -8 to $+7$. The representation of the decimal numbers in 4-bit two's complement codes is shown in Table 6.2.

	Decimal number	Two's complement
Table 6.2 Two's complement codes for decimal numbers	+7	0111
	+6	0110
	+5	0101
	+4	0100
	+3	0011
	+2	0010
	+1	0001
	+0	0000
	−1	1111
	−2	1110
	−3	1101
	−4	1100
	−5	1011
	−6	1010
	−7	1001
	−8	1000

6.3.5.1   Converting Two's Complement Code to Decimal Number

It could be observed in Table 6.2 that the MSB of the two's complement codes of positive decimal numbers is 0 and that of the negative numbers is 1. The equivalent decimal number of two's complement codes is obtained by appropriate decoding operations. Consider the two's complement code, 0101. The MSB, 0, indicates that the equivalent decimal number is positive. The bit is discarded and the equivalent decimal number of 101 is $+5_{10}$.

For the two's complement code, 1011, the MSB, 1, indicates that the equivalent decimal number is negative. The MSB bit is discarded. The remaining bits of the binary number are first complemented and 1 is added to the complemented code. Complement of 011 is 100. Adding 1 to the complemented code gives 101. The equivalent magnitude of decimal number of 101 is 5_{10}. The equivalent decimal number of 1011 is -5_{10}.

6.3.5.2   Advantages

The decimal, 0, has only one code in two's complement method. The roll over feature of one's compliment code is available in two's complement code also. Two's complement codes facilitate the use of same parallel adder circuit for performing addition and subtraction. The arithmetic operations are illustrated with examples.

6.4   Arithmetic Operations with Two's Complement Codes

Consider two decimal numbers. Both the numbers could be positive. One or both the numbers could be negative. Addition or subtraction is performed on the numbers and appropriate sign is assigned to the result in decimal arithmetic. Only addition is performed in binary arithmetic for the two's complement codes of the decimal numbers.

Both positive and negative decimal numbers are encoded into two's complement codes. The two's complement codes of the decimal numbers are treated as unsigned binary numbers and they are added as per the rules of binary addition. Final carry, if any after the addition, is discarded. The sum after discarding final carry is in two's complement code. It is decoded to obtain the equivalent positive and negative decimal numbers. There are four possibilities of binary arithmetic operations using two's complement codes. They are illustrated with examples. The equivalent decimal numbers are also shown in the illustrations.

6.4.1   Illustrations

The four possibilities of arithmetic operations are illustrated with 4-bit two's complement codes in Fig. 6.7a–d. The decimal numbers for arithmetic operations, two's complement codes for the decimal numbers, the results of addition in the codes and the equivalent decimal numbers for the results are shown in the figures. Table 6.2 is used for obtaining the equivalent two's complement codes for decimal numbers and vice versa. Parallel adder is used for performing binary addition and subtraction using two's complement code.

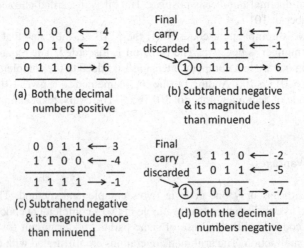

Fig. 6.7 Arithmetic operations with two's complement codes

6.4.2 Parallel Adder for Addition and Subtraction

Let A and B be the decimal numbers. The 4-bit parallel adder circuit arrangement for adding $(A + B)$ or subtracting $(A - B)$ the decimal numbers is shown in Fig. 6.8. The two's complement codes of $+A$ and $+B$ are obtained even if B is negative. The two's complement code of the decimal number, $+A$, i.e. $A_3A_2A_1A_0$ is loaded to the full adders directly. The two's complement code of the decimal number, $+B$, i.e. $B_3B_2B_1B_0$ is loaded to the full adders through XOR gate circuit. The operation of 4-bit adder/subtractor is illustrated with examples. If both A and B are negative decimal numbers, additional gate circuit is needed to obtain the two's complement code of -A before addition.

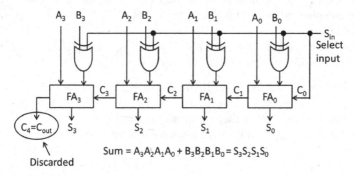

Fig. 6.8 Parallel adder/subtractor using 2's complement codes

6.4.2.1 Addition

Let A is equal to 4 and B is equal to 2. S_{in} is set to 0 for finding the sum, $(A + B)$. The 4-bit two's complement codes of the numbers are:

Two's complement code of $+A$ i.e. $+4 = A_3A_2A_1A_0 = 0100$
Two's complement code of $+B$ i.e. $+2 = B_3B_2B_1B_0 = 0010$

As the select input, S_{in}, is 0, the carry input, C_0, to the LSB full adder, FA_0 is 0. One of the inputs to XOR gates is also 0. Hence, the XOR gates do not complement the bits of B. The bits of B are applied to the full adders without any inversion. The output of the 4-bit parallel adder is the sum of the two's complement of A and B. The final carry out from the addition is 0. The resultant sum is in two's complement code.

Sum $= S_3S_2S_1S_0 = A_3A_2A_1A_0 + B_3B_2B_1B_0 = 0100 + 0010 = 0110$
Equivalent decimal number of $0110 = +6$.

6.4.2.2   Subtraction

Let A is equal to 4 and B is equal to -2. S_{in} is set to 1 for finding the difference, $(A - B)$. The 4-bit two's complement codes of the numbers ignoring the sign of B are:

Two's complement code of $+A$ i.e. $+4 = A_3A_2A_1A_0 = 0100$
Two's complement code of $+B$ i.e. $+2 = B_3B_2B_1B_0 = 0010$

As the select input, S_{in}, is 1, the carry input, C_0, to the LSB full adder, FA_0 is 1. One of the inputs to XOR gates is also 1. Hence, the XOR gates complement the bits of $+B$ i.e. $+2$ and outputs the one's complement of (-2) i.e. 1101. The one's complement of (-2) is applied to the full adders. Adding C_0 i.e. 1 with the LSB of the one's complement of (-2) is equivalent to obtaining the two's complement of (-2) i.e. 1110.

The 4-bit parallel adder outputs the sum of the two's complement codes of $(+4)$ and (-2). The resultant sum is in two's complement code.

Sum $= S_3S_2S_1S_0 = A_3A_2A_1A_0 + (B'_3B'_2B'_1B'_0 + 1)$
Sum $= S_3S_2S_1S_0 = 0100 + (1101 + 1)$
Sum $= 0100 + 1110 = 0010$ after ignoring the carry bit
Equivalent decimal number of two's complement code, $0010 = 2$.

6.4.3   Overflow Detection and Correction

Consider two negative decimal numbers, -7 and -5 for arithmetic addition. The expected result is -12. The addition is performed using 4-bit two's complement codes and the result is shown in Fig. 6.9a. The resulting sum in two's complement code is 0100 after discarding carry. The MSB of the sum is 0 and it indicates that the equivalent decimal number of the sum is positive i.e. $+4$. It is clear that the result is not correct. Overflow is said to have occurred in adder circuit causing erroneous output.

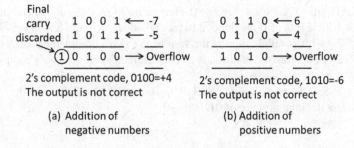

Fig. 6.9 Overflow in the addition of two's complement codes

Consider another example. The addition of two positive numbers, 6 and 4, are shown in Fig. 6.9b using 4-bit two's complement codes. The resulting sum in two's complement code is 1010 i.e. −6. The result is not correct and it is caused by overflow. In general, if the result of addition is outside the range of 4-bit two's complement code (−8 to +7), overflow occurs. Overflow in the addition of two's complement codes could occur with or without final carry output.

6.4.3.1 Detecting Overflow

The schematic diagram of 4-bit parallel adder circuit with XOR gate is shown in Fig. 6.10. Overflow is detected by connecting additional XOR gate to the carry outputs, C_3 and C_4. The output status (F) of the XOR gate indicates overflow in the addition of two's complement codes. If F is 1, overflow has occurred. If F is 0, overflow has not occurred.

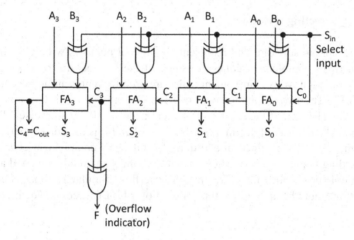

Fig. 6.10 Overflow indicator for 4-bit parallel adder

6.4.3.2 Illustrating Overflow Detection

The two examples in Fig. 6.9 are re-drawn as Fig. 6.11 for understanding overflow detection. $C_3 \oplus C_4$ is equal to 1 in both the examples, indicating overflow in the addition of two's complement codes.

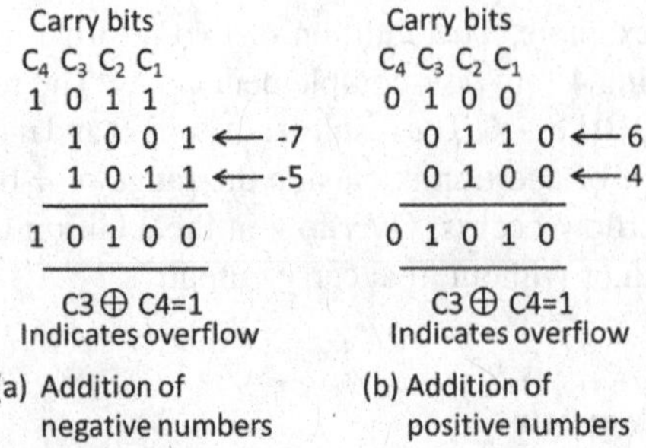

```
    Carry bits            Carry bits
   C₄ C₃ C₂ C₁          C₄ C₃ C₂ C₁
    1  0  1  1           0  1  0  0
       1  0  0  1 ← -7      0  1  1  0 ← 6
       1  0  1  1 ← -5      0  1  0  0 ← 4
    ──────────────        ──────────────
    1  0  1  0  0        0  1  0  1  0
```

$C3 \oplus C4 = 1$
Indicates overflow

$C3 \oplus C4 = 1$
Indicates overflow

(a) Addition of
negative numbers

(b) Addition of
positive numbers

Fig. 6.11 Examples: detection of overflow

The occurrence of overflow could be verified for the four examples in Fig. 6.7 in the addition of two's complement codes. $C_3 \oplus C_4$ is equal to 0 i.e. F is equal to 0, indicating no overflow in all the examples.

6.4.3.3 Correcting Overflow

Using appropriate software and representing decimal numbers with more number of bits in two's complement codes are some of the methods for correcting overflow. Assume 5-bit two's complement codes are used for representing the examples in Fig. 6.11. The two's complement code for -7 and -5 are 11001 and 11011 respectively. The sum of the two's complement codes is 10100 after discarding the final carry. The equivalent decimal number is -12 and it is the correct output. The same sum, 10100, is obtained in addition of 4-bit two's complement codes without discarding final carry. If overflow is detected, the sum could be treated as 5-bit two's complement codes for correcting the overflow. Similar method of correcting overflow could be applied for the addition of 4-bit two's complement codes of 6 and 4.

6.5 Binary Multiplication and Division

Recursive algorithms are available for performing binary multiplication and division. The algorithms are implemented using sequential logic circuits, discussed in subsequent chapters. Multiplication and division are illustrated for 4-bit unsigned binary numbers. The procedure is applicable for two's complement codes also.

6.5.1   Multiplication

Let the 4-bit binary numbers be A ($A_3A_2A_1A_0$) and B ($B_3B_2B_1B_0$). It is assumed that A is multiplicand and B is multiplier. Multiplication, Addition and shift-right operations are performed recursively starting from the LSB (B_0) of multiplier until the operations are performed on MSB (B_3). The intermediate results obtained before the operations on MSB are called partial products. The final result is obtained after performing the operations on MSB. The recursive operations are illustrated in Fig. 6.12.

```
              Multiplicand (A):    0  1  1  0  (6₁₀)
                Multiplier (B):    0  1  0  1  (5₁₀)

Initialize Partial Product:     0  0  0  0
              Add (B₀ * A):  +  0  1  1  0
           Partial Product:     0  1  1  0

                Shift right:    0  0  1  1  0
              Add (B₁ * A):  +  0  0  0  0
           Partial Product:     0  0  1  1  0

                Shift right:    0  0  0  1  1  0
              Add (B₂ * A):  +  0  1  1  0
           Partial Product:     0  1  1  1  1  0

                Shift right:    0  0  1  1  1  1  0
              Add (B₃ * A):  +  0  0  0  0
              Final Product:    0  0  1  1  1  1  0  (30₁₀)
```

Fig. 6.12 Multiplication of binary numbers

The unsigned binary numbers, assigned to A and B, are:

$A = A_3A_2A_1A_0 = 0110 \ (6_{10})$
$B = B_3B_2B_1B_0 = 0101 \ (5_{10})$.

6.5.2   Division

Binary division is similar to the division of decimal numbers. Let the 4-bit binary numbers be A ($A_3A_2A_1A_0$) and B ($B_3B_2B_1B_0$). It is assumed that A is dividend and B is divisor. The bits of dividend starting from MSB are grouped such that the remainder is positive after subtracting divisor from the group of bits. Quotient is 1 if subtraction could be performed and it is 0 if subtraction is not possible. The result of subtraction is called partial remainder. The recursive operation is continued until subtraction is not possible. The division process is completed and the result is the final remainder. Quotient and final remainder are the outputs of binary division. The recursive operations are illustrated in Fig. 6.13.

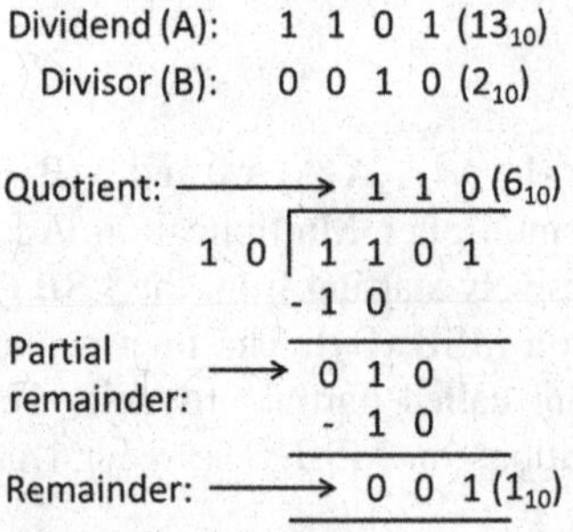

$$\text{Dividend (A):} \quad 1 \ 1 \ 0 \ 1 \ (13_{10})$$
$$\text{Divisor (B):} \quad 0 \ 0 \ 1 \ 0 \ (2_{10})$$

Quotient: $\longrightarrow$ 1 1 0 (6_{10})

$$1 \ 0 \ \overline{\left) 1 \ 1 \ 0 \ 1 \right.}$$
$$\underline{- 1 \ 0}$$

Partial $\longrightarrow$ 0 1 0
remainder:
$$\underline{- \ 1 \ 0}$$

Remainder: $\longrightarrow$ 0 0 1 (1_{10})

Fig. 6.13 Division of binary numbers

The unsigned binary numbers, assigned to A and B, are:

$A = A_3A_2A_1A_0 = 1110 \ (13_{10})$
$B = B_3B_2B_1B_0 = 0101 \ (5_{10})$.

References

1. Walt K (ed) (2004) Analog-digital conversion, analog devices, Mar 2004
2. LaMeres BJ (2017) Introduction to logic circuits & logic design with VHDL. Springer

Chapter 7
Clock and Timing Signals

Abstract Clock signal network exists in computers and digital signal processing systems for controlling and performing their intended functions. A clock signal network has a high stability reference clock generator at the top of the network. The network has other clock generators and timing circuits. The operation of reference clock generator, Schmitt trigger, timer IC 555, non-retriggerable and retriggerable monostable multivibrators are presented.

7.1 Introduction

General purpose clocks regulate practically all our activities. Both 12-h and 24-h clocks are in usage. Let us examine how train traffic is regulated by a clock. Assume that a train is scheduled for departure at 1700 h. When the clock strikes 1700 h, the train does not depart automatically. The clock functions as a reference signal source and it initiates a network of signals for the departure of the train. The train departs when the final green signal is received from the signal network. Similar clock signal network exists in computers and digital signal processing systems for controlling and performing their intended functions.

7.1.1 Clock Signal Network

Basically, a clock signal network has a high stability crystal oscillator circuit at the top of the network and the circuit generates reference clock signal. The clock signal network has additional clock generators and timing circuits. The timing circuits generate a set of clock signals having appropriate characteristics (frequency, duty cycle and delays). The functions of various digital circuits are implemented by the timing circuit signals.

© Springer Nature Switzerland AG 2020

D. Natarajan, *Fundamentals of Digital Electronics*,

Lecture Notes in Electrical Engineering 623,

https://doi.org/10.1007/978-3-030-36196-9_7

7.2 Quality Requirements of Clock Signals

The quality requirements of clock signals are rise time, fall time and jitter. The
requirements are applicable for reference clock signal and other timing clock signals (or simply timing signals). Maximum limits are specified for the requirements
of clock signal in the datasheets of clock ICs. Section 1.2.1 could be referred for
details on rise and fall times of clock signals. Definition and types of clock jitter are
presented.

7.2.1   Clock Jitter

Jitter is the short term phase variation of the significant instants of a digital signal
from their ideal positions in time [1]. Jitter in clock signal occurs due to noise and
interferences. Power supply filtering reduces jitter. Jitter in clock signals is expressed
in units of time.

   As an example, 100 MHz ideal clock signal is shown in Fig. 7.1a. The ideal clock
signal has zero jitter. Clock signals with negligible and significant jitters are shown
in Fig. 7.1b, c respectively.

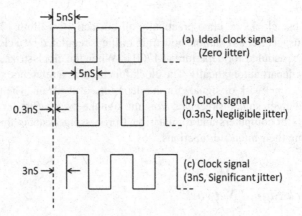

Fig. 7.1 Clock signals with jitter

7.2.1.1   Effect of Clock Jitter

Data and clock signals are the two inputs for many sequential logic devices. Clock
jitter timing limits are specified in the datasheets of clock generator ICs. The specifications are used for designing proper set-up and hold times for data input to devices
to ensure correct data output from the devices. Clock jitter reduces set-up and hold

times of data inputs. For example, if a microprocessor has a data hold time requirement of 2 ns but now the clock jitter is 1.5 ns, then the data hold time is effectively reduced to 0.5 ns, presenting incorrect data to the microprocessor [2]. Set-up and hold times are defined and explained in Chap. 8.

7.2.1.2   Types of Clock Jitter

With ever increasing clock speeds, various measures for clock jitter have been evolved. The types of clock jitter, mentioned in the datasheets of clock ICs are random jitter, period jitter, cycle-to-cycle jitter and phase jitter. Jitters in clock signals are typically caused by noise or other disturbances in the system. Contributing factors include thermal noise, power supply variations, loading conditions, device noise, and interference coupled from nearby circuits [2].

Random jitter is caused by the accumulation of a huge number of processes that each has very small magnitudes; things like thermal noise, variations in trace width, shot noise, etc.; the central limit theorem of probability and statistics gives us a handle on how to describe random jitter distribution [3].

Period jitter is defined in JEDEC Standard 65 as the deviation in cycle time of a clock signal with respect to the ideal period over a number of randomly selected cycles. The data sheets of clock ICs specify the jitter for 10,000 cycles as per the standard.

Cycle-to-cycle jitter is defined in JEDEC Standard 65 as the variation in cycle time of a signal between adjacent cycles, over a random sample of adjacent cycle pairs. The data sheets of clock ICs specify the jitter for 10,000 cycles as per the standard. Application notes of Silicon Labs [4], SiTime™ [2] and seminar notes of Agilent Technologies [3] could be referred for characterizing jitter in clock signals including phase jitter.

7.3   Generating Clock and Timing Signals

High stability crystal oscillator is used for generating reference clock signal. Additional clock generators and timing signals are used considering the performance needs of computer systems. They are designed to function in synchronous or asynchronous with reference clock signal. Synchronous means that logic state transitions in timing signals occur as per the logic state transitions in reference clock signals. Asynchronous timing signals change logic states independently and they are 'free-running' clock signals. Most applications require synchronous clock and timing signals. Reset operations are used in computer systems and the operations are asynchronous with the reference clock signal. If required, free running clock and timing signals could be synchronized with reference crystal oscillator by using appropriate phase locked loop (PLL) and other filter circuits.

Two methods of generating reference clock signal and four types of clock generator circuits using standard gates, Schmitt trigger, 555 Timer IC and monostable multivibrator are presented in this chapter. Monostable multivibrator is also used for generating timing signals for applications requiring time delay.

7.3.1 Reference Crystal Oscillator

Quartz crystal oscillator circuit is used for generating reference clock signal. Quartz crystal is used with ICs having internal oscillator and phase-locked loop (PLL) for generating reference clock signal. Quartz crystals are characterized by low phase noise as they have high Q-factor. However, the frequency of quartz crystals changes with temperature.

Alternatively, temperature controlled crystal oscillators (TCXOs) with additional PLL, buffer and other circuits are also used to generate reference clock signal for high stability applications. TCXOs have ultra-high frequency stability and they are readily available.

7.3.2 Clock Generator Using Standard Gates

Clock signals for simple applications are generated with standard inverter gates (IC 7404) and crystals. An example of schematic diagram is shown in Fig. 7.2. The datasheet of 7404 could be referred for V_{CC} and ground connections for the clock circuit. Schmitt trigger (IC 7414) circuit with crystal is also used for generating clock signal. Application notes [5] of device manufacturers could be referred for additional schematic diagrams for generating clock signals.

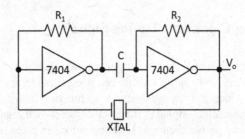

Fig. 7.2 Clock generator using inverters

7.4 Schmitt Trigger

The functional block diagram of Schmitt trigger circuit is shown in Fig. 7.3. The input signal is V_i and the output signal (V_o) is clock signal. Schmitt trigger circuit generates clock signal having sharp rise and fall characteristics from slow rising or noisy periodic input signal. The operation of Schmitt trigger is explained.

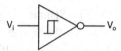

Fig. 7.3 Functional block diagram of Schmitt trigger

7.4.1 Operation

The operation of Schmitt trigger, 5 V TTL IC 74LS14, is explained assuming the input signal is a noisy sine wave. The input noisy sine wave and the output clock signal are shown in Fig. 7.4. When the sinusoidal signal input level is zero or negative, the output of Schmitt trigger is at logic High i.e. at 3.4 V. When the sinusoidal input voltage rises to the specified positive-going threshold voltage, 1.7 V (V_{T+}), the output of Schmitt trigger changes to logic Low, 0.2 V. The output of the Schmitt trigger remains at logic Low (0.2 V) until the sinusoidal signal input voltage level falls to specified negative-going threshold voltage, 0.9 V (V_{T-}). When the sinusoidal signal input voltage level falls to 0.9 V (V_{T-}), the output of the Schmitt trigger changes to logic High (3.4 V). The output clock signal of Schmitt trigger repeats with input sinusoidal signal.

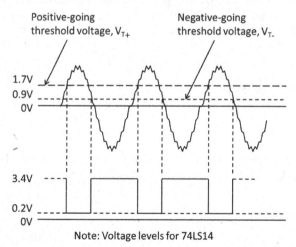

Note: Voltage levels for 74LS14

Fig. 7.4 Input and output waveforms of Schmitt trigger

7.4.2 *Hysteresis*

Schmitt trigger outputs clock signal with sharp rise and fall times with noisy input voltage provided that the noise voltage level variation at any instant of time is less than the hysteresis voltage of the Schmitt trigger. The difference between V_{T+} and V_{T-} is the hysteresis voltage (or simply hysteresis). The hysteresis is the transfer characteristic of Schmitt trigger.

The transfer characteristic of the TTL Schmitt trigger, 74LS14, is shown graphically in Fig. 7.5. The logic state changes at positive-going and negative-going threshold voltages are shown by arrows in the figure. The hysteresis voltage is the difference between the threshold voltages and it is equal to 0.8 V. If the noise voltage level variation of input sinusoidal signal exceeds 0.8 V at any instant of time, multiple transitions would occur at the output of Schmitt trigger.

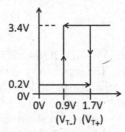

Hysteresis = (1.7 - 0.9) = 0.8V

Note: Voltage levels for 74LS14

Fig. 7.5 Transfer characteristic of Schmitt trigger

CMOS Schmitt trigger ICs could be operated at higher voltages. They have higher noise immunity as the hysteresis voltage level is higher than TTL ICs. The datasheet of the CMOS Schmitt trigger, CD40106B could be referred for operating, output, threshold and hysteresis voltage levels.

7.5   Timer IC, 555

Signetics 555 Timer (1971) is one of the twenty five ICs that shook the world [6]. The IC is active and it is used for generating clock signals. 555 timer IC is used for astable and monostable operations. The operations and the applications of 555 Timer are explained.

7.5.1 Astable Operation

The output of 555 Timer IC changes its logic state from High to Low and from Low to High continuously in astable operation. Timer 555 does not have stable output state. The continuous logic state changes are triggered by R-C network, connected to the IC. Timer 555 is used for generating free running clock signal. The frequency and duty cycle are set by the R-C network. The connection diagram for astable operation is shown in Fig. 7.6.

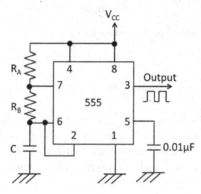

Fig. 7.6 Timer 555: Astable operation

7.5.1.1   Operation

The trigger voltage for the operation of the timer, 555, is applied to the Pin-6 of the IC. The triggering voltage levels are controlled by the voltage level of the capacitor, C. Initially, the charge in the capacitor, C, is zero and the output is at 0 V. When V_{CC} is applied to 555, the capacitor starts charging through the resistors, R_A and R_B. When the capacitor voltage reaches 1/3 of V_{CC}, the output goes to logic High. The output remains at logic High until the capacitor charges to 2/3 of V_{CC}. The capacitor voltage is the trigger voltage. When the trigger voltage reaches 2/3 of V_{CC}, the internal transistor of 555 turns on and it short-circuits the series circuit, (C and R_B) to ground. The capacitor starts discharging through the resistor and simultaneously the output state changes to logic Low. The Low state is maintained until the capacitor discharges to 1/3 of V_{CC}. When the capacitor discharges to 1/3 of V_{CC}, the internal transistor is turned off and the capacitor, C, starts charging again. The output also changes to logic High state. The timer IC 555 triggers itself by the charging and discharging of the capacitor. It functions as a free running clock generator. The trigger waveform at pin-2 and the output clock signal of 555 are shown in Fig. 7.7.

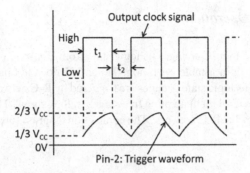

Fig. 7.7 Astable triggering and output waveforms for Timer, 555

7.5.1.2   General Design Computations

The charge time (logic High) is t_1 and it is a function of R_A, R_B and C. The discharge time (logic Low) is t_2 and it is a function of R_B and C. Although the Timer 555 could operate at any voltage (V_{CC}) between 5 and 15 V, t_1 and t_2 are independent of V_{CC}. Timer 555 could be designed for timings from microseconds to hours. The expressions that are useful for design are listed below.

$$t_1 = 0.693(R_A + R_B)C$$

$$t_2 = 0.693 R_B C$$

$$T = t_1 + t_2 = 0.693(R_A + 2R_B)C$$

$$f = \frac{1}{T} = \frac{1.44}{(R_A + 2R_B)C}$$

$$Duty\ cycle = \frac{t_1}{(t_1 + t_2)} = \frac{(R_A + R_B)}{(R_A + 2R_B)}$$

7.5.2   Monostable Operation

Timer 555 IC is used for generating time delays in the monostable mode of operation. The connection diagram for monostable operation is shown in Fig. 7.8. The monostable mode operation of the IC has one stable output state, which is logic Low. The output changes to logic High when an external trigger input pulse is applied to the pin-2 of the IC. The logic High state is a quasi-stable state. The output returns to logic Low after the pre-determined duration.

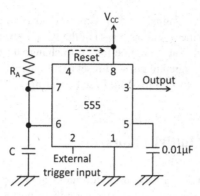

Fig. 7.8 Timer 555: Monostable operation

7.5.2.1 Operation

Normally, the internal transistor of Timer IC 555 is turned on. The external capacitor, C, is short-circuited to ground by the transistor. The output state of 555 Timer remains at logic Low. When an external trigger input is applied to the pin-2 of the IC, the transistor is turned off and the output state changes to logic High. The output change occurs during the negative-going (falling edge) of the trigger pulse. The capacitor also starts charging through the resistor, R_A. When the voltage of the capacitor reaches 2/3 V_{CC}, the internal transistor is turned on. The output state of 555Timer returns to logic Low and the capacitor is short-circuited to ground by the transistor. The operation repeats when another external trigger input is applied to the pin-2 of the IC. The trigger waveform at pin-2, the voltage variation across the capacitor and the output waveform of 555 Timer are shown in Fig. 7.9. The duration of logic High in the output is time delay, t_d.

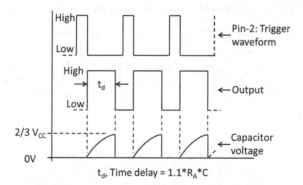

Fig. 7.9 Monostable triggering and output waveforms for Timer, 555

$$t_d = 1.1 R_A C$$

The logic High output signal of 555 Timer could be terminated by applying negative-going pulse to the reset pin (pin-4) of the IC. For example, if td is designed as 10 ms, the logic High state could be forced to change to logic Low state before 10 ms by applying reset signal to pin-4. If reset requirement does not exist, the reset pin should be connected to V_{CC} to avoid false triggering.

7.6   Monostable Multivibrators

Although Timer 555 IC could be used for monostable operation, standard ICs are available as non-retriggerable and retriggerable monostable multivibrators with added features. The devices are known as one-shot devices. ICs 74121 and 74123 are examples of devices for non-retriggerable and retriggerable monostable multivibrators respectively. The ICs have Schmitt trigger inputs to output jitter-free signals. The operations and the applications of the ICs are presented.

7.6.1   Non-retriggerable Monostable Multivibrator

Non-retriggerable monostable devices are used for generating pre-determined time delay signals. The device, 74121, is capable of generating output signal with pulse width (t_w) from 30 ns to 28 s. The pulse width (t_w) is decided by the external R-C network of the device. The device has provisions for active-High and active-Low triggering inputs. The active-High input trigger and the output waveforms are shown in Fig. 7.10. When the output is at logic Low, the device is in stable state. When the output is at logic High, the device is in quasi-stable state. Like the Timer 555, the quasi-state of the output pulse could be forced to its normal state by applying trigger signal to the reset pin of non-retriggerable monostable devices.

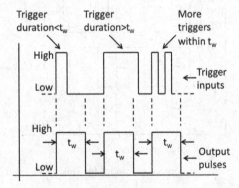

Fig. 7.10 Waveforms of non-retriggerable monostable multivibrator

The device (74121) is called non-retriggerable as its output pulse width (t_w) cannot be altered by retriggering the device when the output is in its quasi-stable state. Even if the trigger pulse duration is higher than t_w, the output pulse width (t_w) cannot be altered. The device could be retriggered only after the output returns to its stable state. The non-triggerable conditions are also shown in Fig. 7.10.

7.6.2 Retriggerable Monostable Multivibrator

IC 74123 is a retriggerable monostable multivibrator. The output pulse width (t_w) of the device could be increased by retriggering the device. As an example, consider the retriggerable monostable multivibrator circuit designed with an output pulse width (t_w), 10 ns. When the output is in quasi-state, assume that a second trigger input pulse is given after 3 ns (t). The quasi-output state will remain for 10 ns from the time at which the second trigger input pulse is given. The output pulse width would be 13 ns (t + t_w). It could be increased further by another retriggering input pulse when the output is still in quasi-state. The trigger input and output pulse waveforms are shown in Fig. 7.11. The retriggerable conditions are also shown in the figure.

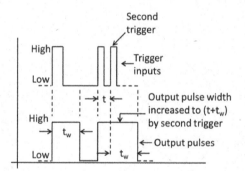

Fig. 7.11 Waveforms of retriggerable monostable multivibrator

Retriggerable monostable multivibrator could be used for detecting rising or falling edge of clock signal as it is sometimes preferable to have a short pulse when an edge is detected rather than sending the original waveform to a microcontroller [7].

References

1. Sullivan DB, Allan DW, Howe Da, Walls FL (1990) NIST technical note 1337. Characterization of clocks and oscillators
2. Clock jitter definitions and measurement methods, Application note SiT-AN10007, SiTime™, Jan 2014
3. Clock jitter analysis with femto-second resolution, Agilent Technologies (2008)

4. Application note on integrated phase noise, AN256 Rev. 0.4 12/11, Silicon Laboratories, USA
5. Williams J (1985) Circuit techniques for clock sources, AN12, Linear Technology, USA
6. Santo BR (2009) 25 microchips that shook the world. IEEE Spectrum, May 2009
7. Maier E (2015) Designing with the SN74LVC1G123 monostable multivibrator, Application report SLVA720, Texas Instruments, USA, July 2015

Chapter 8
Latches and Flip-Flops

Abstract Latches and flip-flops are basic sequential logic devices. The operation of SR and D latches are presented. The application SR latch for eliminating contact bounces in mechanical switches is explained. The operation of D, T and JK flip-flops is explained. Flip-flops using Master-Slave latches are also explained. Set-up time and hold time are the timing requirements of flip-flops. The timing requirements of flip-flop are defined and the method of measuring them is presented.

8.1 Introduction

Basic gates, multiplexer and decoders are examples of combinational logic devices. The output state of the devices depends only on input logic states only. Latches and flip-flops are basic sequential logic devices. The output state of the devices depends on both the input logic states to the devices and the output states of the devices before the application of the inputs.

Latches and flip-flops have two stable (bi-stable) output states, 0 and 1. The output state of the devices remains at 0 or 1 until it is changed by input signals. This feature of a latch or flip-flop is equivalent to storing one bit of data. Storage capability is accomplished through feedback in the devices. The operation of various types of latches and flip-flops is presented. Typical waveforms of input and clock signals are used in timing diagrams for explaining the operation of the devices.

8.2 Latches

Latches are used to "latch onto" information and hold in place [1]. Latches are sensitive to the level (High and Low) of inputs similar to combinational logic devices. The output state of the devices changes as per the level of input signals or latch onto its previous output state as appropriate. The operation of SR latch, $S'R'$ latch, gated

© Springer Nature Switzerland AG 2020
D. Natarajan, *Fundamentals of Digital Electronics*,
Lecture Notes in Electrical Engineering 623,
https://doi.org/10.1007/978-3-030-36196-9_8

SR latch and D latch is presented. The application of SR latches is limited as the devices have forbidden state. D latches are popular for various applications. They are used in servers, printers, buffer registers and are used with multiplexers for parallel to serial data conversion [2].

8.2.1   SR Latch with NOR Gates

Set-Reset (SR) latch is constructed by cross-coupling two NOR gates. The circuit diagram of SR latch and its symbol are shown in Fig. 8.1a, b. The inputs of SR latch labelled as S and R. The outputs are Q and its complement, Q'. The output of SR flip-flop changes as per the input states of S and R without any external control signal. The flip-flop is an asynchronous device. The function table of SR latch is shown in Fig. 8.1c and it is used for explaining the operation of the device. The terminology, function table, is used instead of truth table as it is more appropriate for latches and flip-flops.

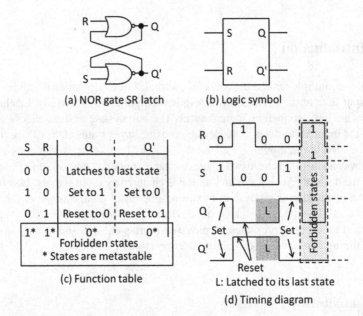

(a) NOR gate SR latch   (b) Logic symbol

S	R	Q	Q'
0	0	Latches to last state	
1	0	Set to 1	Set to 0
0	1	Reset to 0	Reset to 1
1*	1*	0*	0*

Forbidden states
* States are metastable

(c) Function table

L: Latched to its last state

(d) Timing diagram

Fig. 8.1 SR latch

8.2.1.1   Operation

The timing diagram of SR latch is shown in Fig. 8.1d. The inputs of S and R are assumed waveforms in the timing diagram. Propagation delay is ignored in the timing diagram to keep it simple. The outputs (Q and Q′) are explained using the function table of SR latch.

When S = 1 and R = 0, the output (Q) of SR latch is set to 1 and Q′ is set to 0. When S = 0 and R = 1, Q is reset to 0 and Q′ is reset to 1. When S = 0 and R = 0, Q latches to the previous output state i.e. 0; Q′ latches to the previous output state i.e. 1. When S = 1 and R = 0 again, Q is set to 1 and Q′ is set to 0.

If S = 1 and R = 1, both the outputs are forced to 0. The feedback loop in the NOR circuit causes unpredictable output states i.e. metastable states. The variation of propagation delays between the NOR gates and other noise factors in the circuit results in stable but unpredictable output states. Hence, the inputs, S = 1 and R = 1, and the outputs are termed as forbidden states.

8.2.2 S′R′ Latch with NAND Gates

The construction and operation of S′R′ latch is similar to SR latch. S′R′ latch is constructed by cross-coupling two NAND gates. The circuit diagram, logic symbol and function table of S′R′ latch are shown in Fig. 8.2. S′R′ flip-flop is also an asynchronous device as its output changes as per the states of S′ and R′ without any external control signal. The operation of S′R′ latch is similar to SR latch. Timing diagram could be obtained by applying the function table of S′R′ latch for the logic signal waveforms for S′ and R′. Standard IC, 74,279, is available for S′R′ latch.

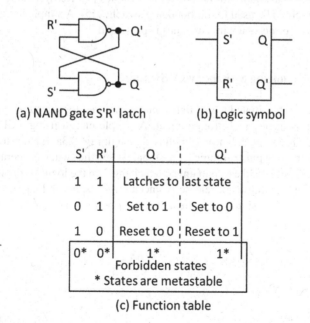

(a) NAND gate S′R′ latch   (b) Logic symbol

S′	R′	Q	Q′
1	1	Latches to last state	
0	1	Set to 1	Set to 0
1	0	Reset to 0	Reset to 1
0*	0*	1*	1*

Forbidden states
* States are metastable

(c) Function table

Fig. 8.2 S′R′ flip-flop

8.2.3   SR Latch for Eliminating Contact Bounce Errors

Generally, digital controls are used for all applications. Mechanical switches are used for various settings and controls. For example, membrane switches are used in home appliances. Micro-switches and push button switches are also used in control panels.

Generally, user actuated mechanical switch operates a logic circuit. When the mechanical switch is actuated, the contact of the switch closes the logic circuit after multiple make-break operations. The multiple make-break operations are called contact bounces. Membranes and springs in mechanical switches cause contact bounces. The contact bounces are interpreted as ones (logic High) and zeroes (logic Low) by the logic circuit. They cause erroneous changes in the output states of the logic circuit.

8.2.3.1   Methods of de-Bouncing Switches

The maximum duration of contact bounces is specified in the datasheets of mechanical switches. Contact bounces in mechanical switches cannot be altered by users. The logic errors caused by the contact bounces of the switches should be eliminated in digital circuits. The elimination of logic errors caused by contact bounces is termed as de-bouncing user operated switches or simply de-bouncing switches.

De-bouncing switches could be achieved by software [3] or by hardware. Standard ICs are also available for de-bouncing switches [4]. Retriggerable monostable multivibrator could be used for de-bouncing switches [5]. A simple hardware circuit using SR latch for de-bouncing switches is presented.

8.2.3.2   De-Bouncing Switches with SR Latch

Assume that a user initiated logic High signal is required for selecting an application in the digital control panel of equipment. A simple circuit using a SPDT (Single Pole Double Throw) push button switch is shown in Fig. 8.3a. Each actuation of the switch generates one pulse of logic High signal. When the switch is operated, operate bounces might occur and cause in unwanted changes in the logic High output signal. When the switch is released, release bounces might occur and cause in unwanted changes in the output signal. Operate and release bounces are shown in Fig. 8.3a in the logic High output signal (Q).

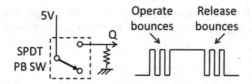

(a) Erroneous output due to contact bounces

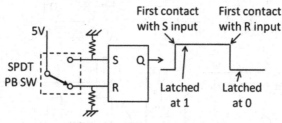

(b) De-bouncing using SR flip-flop

SPDT: Single Pole Double Throw
PB SW: Push Button Switch

Fig. 8.3 Eliminating logic errors of contact bounces

The schematic diagram for de-bouncing push button switch using SR flip-flop is shown in Fig. 8.3b. The pole of the SPDT switch is connected to 5 V. The normally closed contact of the switch is connected to the R input of the flip-flop. The normally open contact is connected to the S input. Initially, $S = 0$ and $R = 1$. Hence the output (Q) of the flip-flop is at 0.

When the push button switch is operated, the pole of the switch makes contact with S input of the flip-flop. S becomes 1 and R becomes 0. Hence, Q changes to 1 for the first contact with S input. When the switch contact opens due to operate bounces, $S = 0$ and $R = 0$. Hence, Q remains at 1 due to latching action of the flip-flop.

When the push button switch is released, the pole of the switch makes contact with R input of the flip-flop. S becomes 0 and R becomes 1 for the first contact with R input. Hence, Q changes to 0. When the switch contact opens due to release bounces, $S = 0$ and $R = 0$. Hence, Q remains at 0 due to latching action. The logic High output of the SR flip-flop is free from unwanted changes in logic states, caused by the contact bounces of the switch.

8.2.4 Gated SR Latch

Gated SR latch is constructed by adding enable input control signal. The control signal enables or disables the inputs, S and R to the latch. The enable control signal could be clock signal. The inputs (S and R) and control signal are routed through

AND gates to SR latch as shown in Fig. 8.4a. The logic symbol and the function table of gated SR latch are shown in Fig. 8.4b, c. Gated SR latch also has forbidden states.

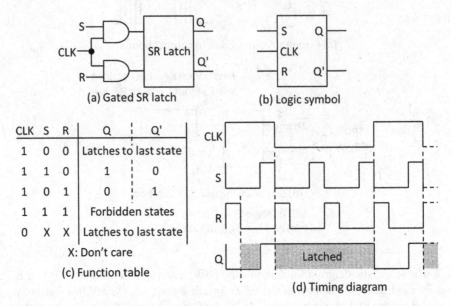

(a) Gated SR latch   (b) Logic symbol

CLK	S	R	Q	Q'
1	0	0	Latches to last state	
1	1	0	1	0
1	0	1	0	1
1	1	1	Forbidden states	
0	X	X	Latches to last state	

X: Don't care

(c) Function table

(d) Timing diagram

Fig. 8.4 Gated SR latch

8.2.4.1 Operation

The timing diagram of gated SR latch is shown in Fig. 8.4d. When the logic status of clock signal is 1, the output state changes or is latched to its previous state as per the inputs (S and R). When the logic status of clock signal is 0, the output state is latched to its previous state.

Initially, CLK = 1, S = 0 and R = 1, Q = 0. When CLK = 1, S = 0 and R = 0, Q latches to previous state, 0. When CLK = 1, S = 1 and R = 0, Q changes to 1. When CLK = 0, S = 0 and R = 1, Q latches to previous state, 1. Q remains latched to 1 until CLK = 1. S and R are don't care inputs when CLK = 0. The output states of the latch could be obtained for the remaining input states.

8.2.5 D Latch

The construction of D latch i.e. Data latch is the extension of gated SR latch. Clock signal is used as enable signal for D latch. The schematic diagram of D latch is shown in Fig. 8.5a. The input, D, is routed through an inverter to the input, S, of gated SR latch and it is directly connected to the input, R. The input circuit of D latch eliminates the input state (S = 0 and R = 0) and the forbidden state (S = 1 and R = 1) of the gated SR latch. The logic symbol and the function table of D latch are shown in Fig. 8.5b, c.

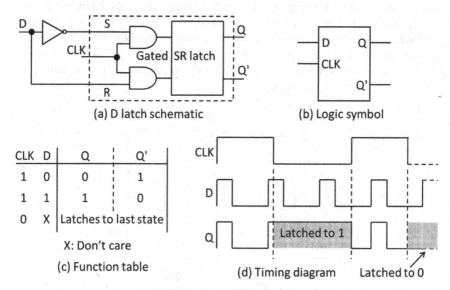

(a) D latch schematic

(b) Logic symbol

CLK	D	Q	Q'
1	0	0	1
1	1	1	0
0	X	Latches to last state	

X: Don't care

(c) Function table

(d) Timing diagram   Latched to 0

Fig. 8.5 D Latch

8.2.5.1 Operation

The timing diagram of D latch is shown in Fig. 8.5d. When CLK = 1, the output state changes as per the input, D. In other words, the latch is said to be transparent to input, D when CLK = 1. When CLK = 0, the output state is latched to its previous state. For example, if CLK = 0 when D = 1, the output state is latched to 1. Standard ICs (Ex. 74HC75) are available for D latches.

8.3 Flip-Flops

Latches are level sensitive devices whereas flip-flops are edge-triggered devices. For example, the output state of D latch changes when clock signal is High as per input signal. The output of flip-flops changes at the rising edge or the falling edge of clock signal as per input signal. Between the edges of clock signal, the inputs to flip-flops are don't care inputs. Flip-flops that use the rising edge i.e. the transition from 0 to 1 of clock signal are positive edge-triggered devices. Flip-flops that use the falling edge i.e. the transition from 1 to 0 of clock signal are negative edge-triggered devices. Edge-triggering ensures that the operation of flip-flops is in synchronous with reference clock signal.

Assume that a flip-flop is positive edge-triggered. The input and output waveforms of positive edge-triggered flip-flop is shown for three clock triggers (Trigger-1, Trgger-2 and Trigger-3) in Fig. 8.6. The positive triggering is shown by the upward arrow in the clock signal. Output state (Q) changes at each clock trigger. Although input data changes between triggers, the output state of flip-flop does not change; it remains latched to its previous state. If required, the output state of flip-flop could be forced to change using reset signal. Similar waveforms could be obtained for negative edge-triggered flip-flops.

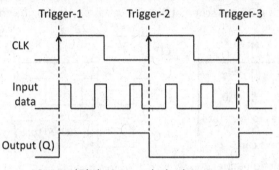

Output (Q) changes at clock edges. Input
data changes between triggers ignored.

Fig. 8.6 Positive edge-triggering of a flip-flop

8.3.1   Function Table

The relationship between the input and output states of flip-flop is shown in function table. The output state (Q) of flip-flop prior to edge-triggering is termed as present state and it is denoted as Q_n in function table. The output state (Q) of flip-flop after edge-triggering is termed as next state and it is denoted as Q_n^+. Next state (Q_n^+) depends on both the inputs and the present state (Q_n) of flip-flop. The output of flip-flop remains at Q_n^+ until the arrival of next clock trigger. It should be noted that the next state (Q_n^+) after clock Trigger-1 is the present state (Q_n) for clock Trigger-2; the next state (Q_n^+) after clock Trigger-2 is the present state (Q_n) for clock Trigger-3 and so on for other clock triggers.

8.3.2 Applications

Flip-flops are used for storing data in memory devices such as registers. They are also used in counters and dividers. The operation of edge-triggered D, T and JK flip-flops are presented. Although edge-triggered SR flip-flop could be constructed, the application of the flip-flop is not popular as the device has forbidden state.

8.3.3 D Flip-Flop

D flip-flop is widely used and it is known as data or delay flop-flop [1]. The logic symbols of positive and negative edge-triggered D flip-flops are shown in Fig. 8.7. Standard ICs (Ex. 74HC74) are available for D flip-flops.

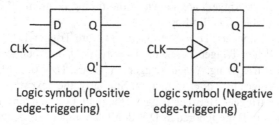

Logic symbol (Positive edge-triggering)   Logic symbol (Negative edge-triggering)

Fig. 8.7 D flip-flops

8.3.3.1 Operation

The function table of positive edge-triggered D flip-flop is shown in Fig. 8.8a. Three clock triggering conditions are shown in the function table. The output states (Q) of D flip-flop for the three triggering conditions are:

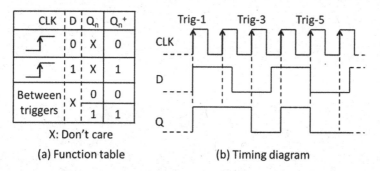

(a) Function table   (b) Timing diagram

Fig. 8.8 Function table and timing diagram of D flip-flop

(i)   Triggering when D = 0:
         If Q_n is 1, Q_n^+ changes to 0; If Q_n is 0, Q_n^+ remains at 0.
         In other words, Q_n is don't care input (X) for the trigger when D = 0.
(ii)   Triggering when D = 1:
         If Q_n is 1, Q_n^+ remains at 1; If Q_n is 0, Q_n^+ changes to 1.
         In other words, Q_n is don't care input (X) for the trigger when D = 1.
(iii) Between triggers
         The output state of the flip-flop does not change for input state changes
between two consecutive triggers i.e. D is don't care input (X).

8.3.3.2   Timing Diagram

The timing diagram of D flip-flop is shown in Fig. 8.8 (b). The outputs (Q) of D
flip-flop for six clock triggers are shown in the timing diagram. They are derived
from the function table of D flip-flop.

When Trigger-1 appears, D = 1. As Q_n is 0 before Trigger-1, Q_n^+ changes to 1
after triggering. When Trigger-2 appears, D = 1. As Q_n is 1 before Trigger-2, Q_n^+
remains at 1 after triggering. When Trigger-3 appears, D = 0. As Q_n is 1 before
Trigger-3, Q_n^+ changes to 0 after triggering. Similarly, the next state, Q_n^+, could be
obtained for other triggers. Between triggers, the output (Q) remains in the previous
state. The changes in D are ignored as D is don't care input between triggers.

8.3.3.3   Comparing D Latch and D Flip-Flop

The timing diagrams of D latch and D flip-flop with identical clock signal and data
input are shown in Fig. 8.9. The output of D latch changes multiple times during a
clock cycle as per the input data when the level of clock signal is 1. The output of D
flip-flop changes once during a clock cycle as per the input data at the positive edge
of clock signal.

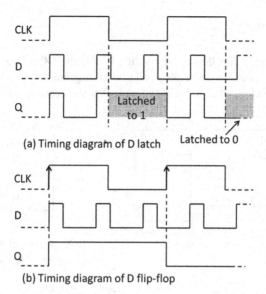

(a) Timing diagram of D latch   Latched to 0

(b) Timing diagram of D flip-flop

Fig. 8.9 Timing diagrams of D latch and D flip-flop

8.3.3.4 Preset and Clear Inputs

When power is applied to digital systems, it is practically impossible to predict the output states of D flip-flops used in the systems. Generally, D flip-flops are initialized to known output states for stable operations. They are initialized to 1 or 0 considering the requirements of applications.

D Flip-flop has two additional inputs, namely, Preset and Clear. Preset input initializes the output of flip-flop to 1. Clear input initializes the output of flip-flop to 0. Preset and Clear inputs are asynchronous inputs. The inputs force the output state of D flip-flop to 1 or 0 as applicable irrespective of the status of clock signal and data input. Standard ICs for D flip-flops use the terms, set and reset, instead of preset and clear. The logic symbol of D flip-flop with the active Low set and reset inputs is shown in Fig. 8.10.

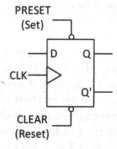

Fig. 8.10 D flip-flop with preset and clear inputs

8.3.4   T Flip-Flop

T flip-flop or Toggle flip-flop is the modified form of D flip-flop. The schematic diagram, logic symbol, function table and timing diagram of T flip-flop is shown in Fig. 8.11. The function table defines Q_n^+ for clock triggers and input states. The input states are of T and Q_n.

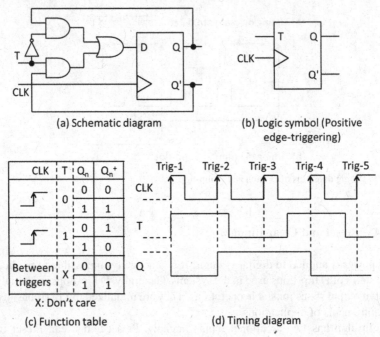

(a) Schematic diagram   (b) Logic symbol (Positive edge-triggering)

X: Don't care

(c) Function table   (d) Timing diagram

Fig. 8.11   T flip-flop

8.3.4.1   Operation

The operation of T flip-flop is similar to that of D flip-flop. The timing diagram of T flip-flop is shown in Fig. 8.11d. The outputs (Q) of T flip-flop for the five clock triggers are shown in the timing diagram. They are obtained by applying the function table of T flip-flop to the input waveforms.

When Trigger-1 appears, T = 1. As Q_n is 1 before Trigger-1, Q_n^+ changes to 0 after triggering. When Trigger-2 appears, T = 1. As Q_n is 0 before Trigger-2, Q_n^+ changes to 1 after triggering. When Trigger-3 appears, T = 0. As Q_n is 1 before Trigger-3, Q_n^+ remains at 1 after triggering. Similarly, the next state, Q_n^+, could be obtained for Trigger-4 and Trigger-5. Between triggers, the output (Q) remains in the previous state. The changes in T are ignored as T is don't care input between triggers. It could be summarized that $Q_n^+ = (Q_n)'$ when T = 1 and $Q_n^+ = Q_n$ when T = 0. Standard ICs for T flip-flop is available with preset and clear inputs.

8.3.4.2 Toggling Output

When the data input of T flip-flop is 1, the output (Q) of the flip-flop toggles between 1 and 0 for multiple edge-trigger inputs. The toggling output of T flip-flop is shown in Fig. 8.12 for five clock triggers. It could be observed that the frequency of the output signal (Q) is half of the frequency of clock signal. The toggling output of T flip-flop is used in counters, dividers and timers.

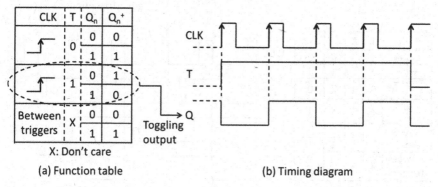

(a) Function table   (b) Timing diagram

Fig. 8.12 Toggling output of T flip-flop

8.3.5 JK Flip-Flop

JK flip-flop is the modified type of D flip-flop with two inputs, J and K. The schematic diagram, logic symbol and function table of JK flip-flop is shown in Fig. 8.13. The function table defines Q_n^+ for clock triggers and input states. The input states are of J, K and Q_n.

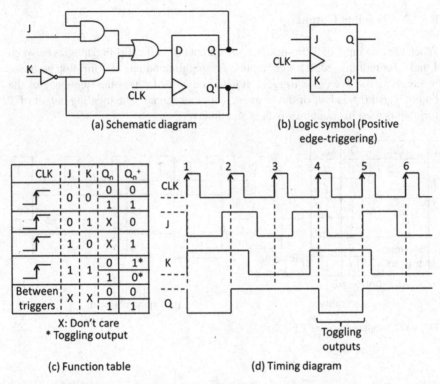

(a) Schematic diagram

(b) Logic symbol (Positive edge-triggering)

X: Don't care
* Toggling output

(c) Function table

(d) Timing diagram

Fig. 8.13 JK flip-flop

Clock signal with six triggers and input data (J and K) are shown in the timing diagram in Fig. 8.13d. The outputs (Q) of JK flip-flop for the six clock triggers are shown in the timing diagram. They are obtained by applying the function table of JK flip-flop to the input waveforms. Toggling outputs (Q) are also shown in the timing diagram.

8.4 Flip-Flop Timing Requirements

Minimum set-up time and hold time for data inputs, maximum clock frequency and minimum clock pulse width are the timing requirements of flip-flops. They are designed to ensure correct output from flip-flops. Limits for the timing requirements are specified in the datasheets of flip-flops by IC manufacturers. Clock frequency and pulse width are not discussed as the specifications are self-explanatory. The requirements of set-up time and hold time for data inputs to flip-flops are explained.

8.4.1 Set-up and Hold Time

The timing diagrams of D, T and JK flip-flops shown in Sec. 8.3 are valid for ideal input and output conditions. Rise time, fall time and propagation delay are assumed to be zero for the timing diagrams of the flip-flops. The timing diagrams show that data input is present at the instant of edge triggering. The ideal conditions do not exist. It is necessary that the data input to flip-flops should be present for sufficient time before and after edge-triggering to ensure correct output from flip-flops.

8.4.1.1 Definition of Set-up and Hold Time

The minimum time for the data input that must be present before edge-triggering is the set-up time (T_{su}) of flip-flop. The minimum time for the data input that must be present after edge-triggering is the hold time (T_h) of flip-flop. T_{su} and T_h are applicable for latches also except that data input should be present before and after enable or clock signal accordingly. Characterizing T_{su} and T_h are explained for D flip-flop. It could be extended for all other types of devices.

8.4.1.2 Characterizing Set-up and Hold Time

T_{su} and T_h for D flip-flop with data input (D) are shown in Fig. 8.14. They are characterized considering the propagation delay of the device. Conventionally, setup and hold times are independently characterized as the setup and hold skews so that the increase in CK-to-Q delay remains within a certain amount of percentage (e.g. 10%) [6]. CK-to-Q delay represents propagation delay in the output (Q) of flip-flop after edge triggering the clock (CK) signal. It is the criterion for deciding set-up (T_{su}) and hold (T_h) time requirements.

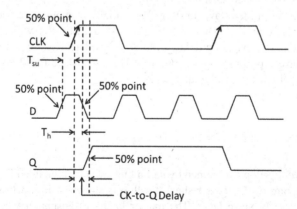

Fig. 8.14 Set-up and hold time requirements for D flip-flop

Initially, the data input signal duration of flip-flop before clock edge triggering is set high. The CK-to-Q delay of device is observed. The data signal duration before clock edge triggering is reduced gradually towards the trigger, monitoring the CK-to-Q delay. The minimum set up time, T_{su}, is the duration of data signal, which causes the increase of CK-to-Q delay and the increase is limited to 10%.

The procedure for measuring T_{su} is applied for measuring T_h also. Initially, the data input signal duration of flip-flop after clock edge triggering is set high. The data signal duration after clock edge triggering is reduced gradually towards the trigger, monitoring the CK-to-Q delay. The minimum hold time, T_h, is the duration of data signal, which causes the increase of CK-to-Q delay and the increase is limited to 10%.

8.5   Flip-Flops Using Master-Slave Latches

Flip-flops could be constructed by cascading two latches. Input data is connected to the first device and the device is called master latch. The output of master latch is connected to the second device and the device is called slave latch. Although the master and slave latches are level-triggered, the final output of master-slave latches is same as that of edge-triggered flip-flop. The final output changes once during a clock cycle as per input data, which is required in the design of many sequential logic circuits.

Master-slave flip-flops are widely used in digital ICs for various application specific requirements. Timing analysis shows that a more efficient and robust design results if combinational logic is placed between the master and slave latches rather than simply abutting them to form a flip-flop [7]. The structure and performance of master-slave flip-flop are improved over a period of time. Pulse based enhanced scan flip-flop has 13% lower power dissipation and 26% better timing than a conventional D flip-flop based enhanced scan flip-flop [8].

Standard ICs for master-slave flip-flops are available with improved performance. MC10H131 is an example for Dual D Type Master-Slave Flip-flop with improvement in clock speed and propagation delay. Similarly, standard ICs are available for JK type master-slave flip-flops. The operation of D master-slave flip-flop is explained assuming ideal clock and data input signals.

8.5.1   D Master-Slave Flip-Flop

D Master-Slave flip-flop is obtained by cascading two clock-enabled D latches. The schematic diagram of the master-slave flip-flop is shown in Fig. 8.15a. The clock signal is inverted for slave latch. The master latch is transparent when the clock signal is positive. When the clock signal is negative, the slave latch is transparent. The negative clock signal is actually converted into positive clock signal by the invertor for the slave latch.

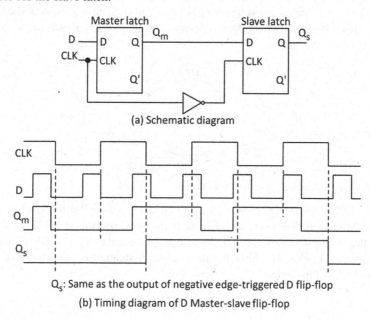

(a) Schematic diagram

Q_s: Same as the output of negative edge-triggered D flip-flop

(b) Timing diagram of D Master-slave flip-flop

Fig. 8.15 D master-slave latch

The timing diagram of D master-slave latch is shown in Fig. 8.15b. D is the data input of master latch and Q_m is its output. The waveforms of clock signal, data input, D, and Q_m are shown in the figure. Q_m is transparent to the data input when the clock signal is positive and it is latched when the clock signal is negative.

The output of master latch is Q_m and it is the input to slave latch. Q_s is the output of slave latch. Q_s is transparent to Q_m when the clock signal is negative and it is latched when the clock signal is positive. It could be observed that the final output waveform of Q_s is same as the output of negative edge-triggered D flip-flop.

8.6   State Transition Diagram of Flip-Flop

Function tables define the relationship between the inputs and output of flip-flops. In other words, they define the functional behavior of flip-flops. Timing diagrams define the functional behavior of flip-flops graphically. State transition diagram is another method of defining the functional behavior of flip-flop. Finite State Machine Modelling (FSM) tool use state transition diagrams in the design of sequential logic circuits. FSM is presented in greater details in Chap. 13.

The state transition diagram of D flip-flop is shown in Fig. 8.16. The output state of flip-flop is either 0 or 1. A circle is used to represent the state of flip-flop. The two states are shown in two circles and the circles are marked as Q = 0 and Q = 1. When the flip-flop is edge-triggered, the output state of the flip-flop could change from 0 to 1 or from 1 to 0. If D = 1 when Q = 0, the output state changes from 0 to 1. If D = 0 when Q = 1, the output state changes from 1 to 0. State transition is shown by curved arrows between circles. The two possible state changes are shown by curved arrows between the circles.

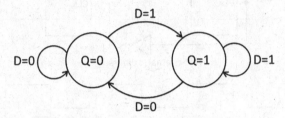

Fig. 8.16 State transition diagram for D flip-flop

When D flip-flop is edge-triggered, the output state of the flip-flop could remain at the same state. If D = 1 when Q = 1, the output state remains at 1. If D = 0 when Q = 0, the output state remains at 0. Remaining in the same states is shown by small arcs around the circles. Similarly, the state transition diagrams could be drawn for T flip-flop and JK flip-flop.

References

1. Modelling Latches and Flip-flops (2013) Lab Workbook, Xilinx
2. Datasheet of Single D-Type Latch with 3-state output, SN74LVC1G373, Texas Instruments, May 2017
3. Eric M (2018) Schwartz, Switch De-bouncing with software, EEL3744C, Rev. 0, University of Florida, June 2018
4. Switch Bounce and Other Dirty Little Secrets, Maxim Application Note 287, Dallas Semiconductor, Sep 2000
5. Maier Emrys (2015) Designing with the SN74LVC1G123 monostable multivibrator, application report SLVA720. Texas Instruments, USA
6. Okumura T, Hashimoto M (2010) Setup time, hold time and clock-to-Q delay computation under dynamic supply noise. In: IEEE custom integrated circuits conference
7. Dally W, Poulton J (2001) Digital system engineering. Cambridge Univ. Press, Cambridge
8. Kumar R, Bollapalli KC, Garg R, Soni T, Khatri SP (2009) A robust pulsed flip-flop and its use in enhanced scan design. In: IEEE international conference on computer design

Chapter 9
Registers

Abstract Registers are used for storing data. Two or more flip-flops are cascaded in registers. n-bit register requires n flip-flops for storing n bits of data. The operation of storage registers, shift registers and universal shift register are presented. Four applications of shift registers are illustrated with examples. The applications are Ring counter, Johnson counter, Linear Feedback Shift Register (LFSR) and Serial adder.

9.1 Introduction

A register is a sequential logic block. Two or more flip-flops are used in registers. A flip-flop is capable of storing one bit of data whereas a register is capable of storing many bits of data. n-bit register requires n flip-flops for storing n bits of data. Common clock signal is used for the flip-flops. In addition to flip-flops, combinational gate circuits are generally used in registers to support the operation of the registers. Like flip-flops, registers are temporary storage devices. The output states of registers remain at 1 or 0 until the states are cleared for storing new bits of data.

9.2 Storage Registers

The operation of simple storage register and standard storage register is explained. Simple storage register is useful for understanding the basic operation of registers. Standard storage registers are used in many applications such as buffer registers, bus drivers and transferring digital data to input-output ports. They have additional controls for loading input data and reading output data. ICs are available for standard storage registers.

© Springer Nature Switzerland AG 2020

D. Natarajan, *Fundamentals of Digital Electronics*,
Lecture Notes in Electrical Engineering 623,
https://doi.org/10.1007/978-3-030-36196-9_9

9.2.1 Simple Storage Register

The schematic diagram of simple 4-bit storage register is shown in Fig. 9.1. Four
positive edge-triggered D flip-flops are used. The flip-flops have common clock
signal. The 4-bit input data to the flip-flops of the register is $D_3D_2D_1D_0$. The input
data is available at the output of the register as $Q_3Q_2Q_1Q_0$ on the next positive
transition of clock input. The inputs and the outputs are shown for each flip-flop in
the figure. The output data is retained i.e. stored by the register until it is read. The
output states of register are reset to 0 prior to triggering for storing the next 4-bit
input data.

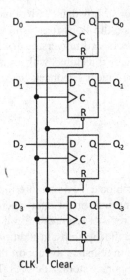

Fig. 9.1 4-bit simple storage register

9.2.2 Standard Storage Register

The simplified schematic diagram of standard 4-bit standard storage register is shown
in Fig. 9.2. The register has two additional controls, namely, Data disable and Output
disable. The schematic diagram of the register is based on the datasheets of ICs
(Ex.: CD4076 and MC14076B) for standard storage registers. The basic operation of
standard storage register is same as that of simple storage register regarding reset and
transferring data from the input ports to the output ports of flip-flops. The operation
of the controls is presented.

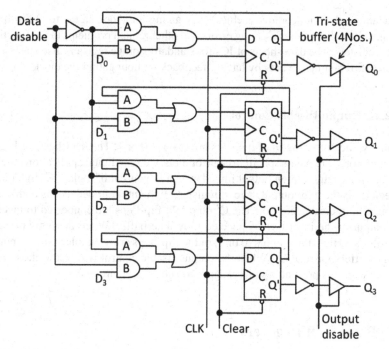

Fig. 9.2 4-bit standard storage register

9.2.2.1 Data Disable Control

Data disable control could be set High or Low for performing two operations. When it is set High, input data is loaded into the flip-flops of register. When it is set Low, the register retains the stored data.

The data disable control circuit of each D flip-flop has two AND gates (A and B) and one OR gate. Normally, the data disable input is Low. When the data disable input is High, the AND gates, B, are selected and data is loaded into the input ports of D flip-flops. The data disable input is set Low after loading the data. Data is transferred to the output of the flip-flops when the next positive transition of clock signal appears.

The output data of register is used by subsequent circuits. Control is necessary for retaining the data at the outputs of flip-flops until it is read. Output data is retained by having feedback loop from the output, Q, of the flip-flops. The feedback loop is connected to the AND gates, A, of the data disable control circuit. The AND gates, A, are selected as the data disable input is low. As the inputs and the outputs of the flip-flops are same, the output data is retained in the flip-flops with clock triggering also.

Retaining the data could be achieved by gating the clock signal to D flip-flops. Gating the clock introduces timing errors and other related problems in the functioning of registers. The data sheets of ICs for standard storage registers also specify the feature as 'retaining the data by having feedback without gating the clock'.

9.2.2.2 Output Disable Control

Normally, the output states of registers are two i.e. 0 or 1. The datasheets of ICs for standard storage registers specify tristate outputs. Certain digital applications such as data bus lines require high output impedance at the output of registers. High output impedance is the third output state of register. High output impedance is realized by having output disable control. The Q' output of flip-flops are connected to inverter. The output of the inverter is connected to tristate buffer. When the output disable control is set High, the outputs (0 or 1) of the flip-flops are available at the outputs of tristate buffers for reading. When the output disable control is set Low, the outputs of tristate buffers present high impedance to subsequent circuits.

9.3   Basic Shift Registers

Flip-flops are cascaded in shift registers. The output data of first flip-flop is the input data for the second flip-flop; the output data of second flip-flop is the input data for the third flip-flop and so on. The flip-flops are edge-triggered using common clock signal for transferring i.e. shifting data from one flip-flop to next flip-flop in shift registers.

There are four basic types of shift registers. The registers are named considering the mode of shifting input data. Serial-in/parallel-out shift register, parallel-in/serial-out shift register, serial-in/serial-out shift register and parallel-in/parallel-out shift register are the four basic types of shift registers. Input data is shifted from left to right in the shift registers.

Standard ICs are available for the basic types of shift registers. The ICs use CLOCK INHIBIT gate circuit to control the shifting of data between the flip-flops of shift registers. The gate circuit disables clock signal to prevent changes in the logic states of flip-flops until the output of shift registers is read. The clock signal is enabled by the gate circuit for shifting new input data. The functioning of the four types of 4-bit shift registers using D flip-flops is presented. CLOCK INHIBIT gate circuit is not shown in the schematic diagrams of the shift registers.

9.3.1 Serial-In/Parallel-Out Shift Register

The simplified schematic diagram of 4-bit Serial-in/Parallel-out shift register is shown in Fig. 9.3. The functioning of the register is illustrated for the 4-bit input data, $A_3A_2A_1A_0$. The binary bits of data are made available bit by bit (serially) to the input of the first flip-flop (FF-1) of the shift register. After the input bits of data are shifted out by the flip-flops, the data bits are available simultaneously (parallel format) at the output of the flip-flops of the shift register.

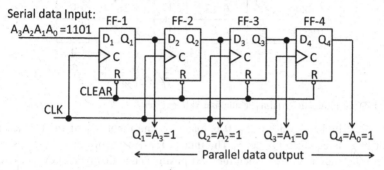

Fig. 9.3 Serial-in/Parallel-out 4-bit Shift register

9.3.1.1 Operation

The output states of the flip-flops of 4-bit Serial-in/Parallel-out Shift register are reset to 0 using Clear control before loading input data to the register. Assume that the four bit word, $A_3A_2A_1A_0$ (1101), is loaded to the flip-flop, FF-1, bit by bit starting from LSB.

The LSB A_0, 1 is loaded to the input, D_1 of FF-1. The inputs, D_2, D_3 and D_4 of other flip-flops are 0. The flip-flops are triggered when the next positive edge (Trigger-1) of clock signal appears. The output state, Q_1, of FF-1 changes to 1. The output states (Q_2, Q_3 and Q_4) of other flip-flops remain at 0 as their inputs are at 0. The output of the shift register in parallel format is 1000 and it is shown in Fig. 9.4.

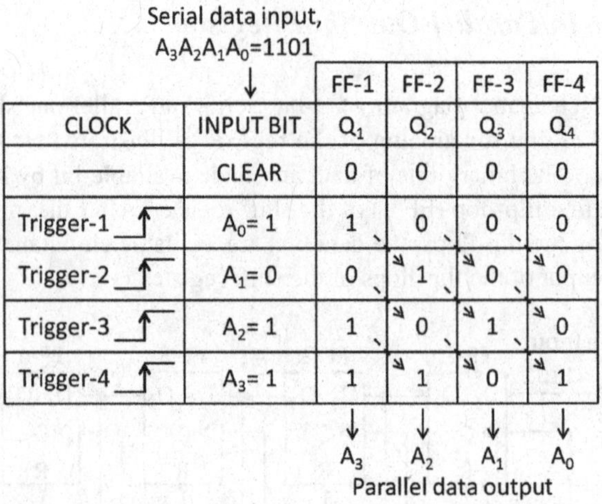

Serial data input,
$A_3A_2A_1A_0 = 1101$

CLOCK	INPUT BIT	FF-1 Q_1	FF-2 Q_2	FF-3 Q_3	FF-4 Q_4
——	CLEAR	0	0	0	0
Trigger-1	$A_0 = 1$	1	0	0	0
Trigger-2	$A_1 = 0$	0	1	0	0
Trigger-3	$A_2 = 1$	1	0	1	0
Trigger-4	$A_3 = 1$	1	1	0	1

A_3 A_2 A_1 A_0
Parallel data output

Fig. 9.4 Data shifting in Serial-in/Parallel-out Shift register

The next higher level bit, A_1, 0 is loaded to the input, D_1 of FF-1. The input, D_2, of FF-2 is equal to Q_1 i.e. 1. The inputs, D_3 and D_4 of other flip-flops are 0. The flip-flops are triggered when the next positive edge (Trigger-2) of clock signal appears. The output state, Q_1, of FF-1 changes to 0. The output state, Q_2, of FF-2 changes to 1. The output states (Q_3 and Q_4) of other flip-flops remain at 0 as their inputs are at 0. The output of the shift register becomes 0100. It could be seen that the output bits of flip-flops after Trigger-1 are right-shifted Trigger-2. Shifting is shown by dashed arrows in the figure.

Similarly, the next higher level bit (A_2) and MSB (A_3) are loaded bit by bit. The changes in the output states of the flip-flops for Trigger-3 and Trigger-4 are shown in Fig. 9.4. The clock signal is inhibited after Trigger-4. The output states of the flip-flops represent the input word, $A_3A_2A_1A_0$. In other words, the serial data input is converted into parallel data output by the shift register.

9.3.2 Parallel-In/Serial-Out Shift Register

Standard ICs for Parallel-in/Serial-out Shift registers (Ex.: 74LS166) have CLOCK INHIBIT and Load/Shift control gate circuits. The simplified schematic diagram of 4-bit Parallel-in/Serial-out Shift register is shown in Fig. 9.5. The bits of input data are loaded to the flip-flops of the shift register simultaneously (parallel format). The last flip-flop (FF-4) of the register outputs the data, bit by bit (serially).

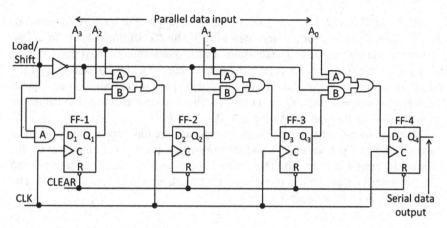

Fig. 9.5 Parallel-in/Serial-out 4-bit Shift register

9.3.2.1 Operation

The output states of the flip-flops of 4-bit Parallel-in/Serial-out Shift register are reset to 0 using Clear control before loading input data to the register. The Load/Shift control is set High for loading the four bit word, 1101 ($A_3A_2A_1A_0$) to the register. The four AND gates, A, of the control gate circuit are activated and the four bits are simultaneously loaded to the flip-flops of the register. The MSB (A_3), 1, is loaded to D_1 of FF-1; the next lower bit (A_2), 1, is loaded to D_2 of FF-2; the next lower bit (A_1), 0, is loaded to D_3 of FF-3 and the LSB (A_0), 1, is loaded to D_4 of FF-4. The inputs and the outputs of the flip-flops after loading are shown in Fig. 9.6.

		Parallel data input ↓	FF-1	FF-2	FF-3	FF-4	
OPERATION	CLOCK	INPUT	Q_1	Q_2	Q_3	Q_4	
CLEAR	——	——	0	0	0	0	
LOAD Load/Shift=1	——	$D_1=A_3=1$ $D_2=A_2=1$ $D_3=A_1=0$ $D_4=A_0=1$	0	0	0	0	Serial data outputs
SHIFT Load/Shift=0	Trigger-1 ↑	——	1	1	0	1	→ A_0
	Trigger-2 ↑	——	0	1	1	0	→ A_1
	Trigger-3 ↑	——	0	0	1	1	→ A_2
	Trigger-4 ↑	——	0	0	0	1	→ A_3

Fig. 9.6 Data shifting in Parallel-in/Serial-out Shift register

The Load/Shift control is set Low for shift operation. The three AND gates, B, of the control gate circuit are activated to facilitate the shifting of bits from one flip-flop to the next flip-flop. The flip-flops are triggered when the next positive edge (Trigger-1) of clock signal appears. The output states (Q_1, Q_2, Q_3 and Q_4), of the flip-flops are shown in Fig. 9.6. The bit at Q_4 of the FF-4 is the LSB of input data for reading. The outputs at Q_1, Q_2 and Q_3 of the flip-flops are transferred to the inputs of next flip-flops through the AND gates, B.

When the next positive edge (Trigger-2) appears, the flip-flops are triggered. The output states of flip-flops change and the next higher bit, A_1, is available at Q_4 for reading by subsequent circuit. The shifting operations continue for Trigger-3 and Trigger-4 until the all the bits of input data are available serially at Q_4 of FF-4. The clock signal is inhibited after Trigger-4.

9.3.3 Serial-In/Serial-Out Shift Register

Standard ICs (Ex. 74LS164) are available for Serial-in/Serial-out Shift Register. The construction of Serial-in/Serial-out Shift Register is similar to Serial-in/Parallel-out Shift Register. The simplified schematic diagram of Serial-in/Serial-out Shift Register is shown in Fig. 9.7. It could be observed that the figure is reproduction of Fig. 9.3 without parallel data output. The bits of input data are loaded to the shift register bit by bit (serially). The last flip-flop (FF-4) of the register outputs the data bit by bit (serially).

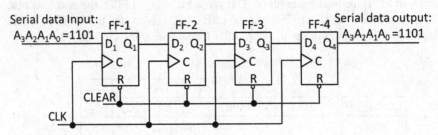

Fig. 9.7 Serial-in/Parallel-out 4-bit Shift register

9.3.3.1 Operation

The output states of the flip-flops of 4-bit Serial-in/Serial-out Shift register are reset to 0 using Clear control before loading input data to the register. The operation of Serial-in/Serial-out Shift register is similar to Serial-in/Parallel-out Shift register except that additional clock triggers are used. The four bit word, $A_3A_2A_1A_0$ (1101) is loaded to the flip-flop, FF-1, bit by bit starting from LSB.

The LSB A_0, 1 is loaded to the input, D_1 of FF-1. The inputs, D_2, D_3 and D_4 of other flip-flops are 0. The flip-flops are triggered when the next positive edge

(Trigger-1) of clock signal appears. The output state, Q_1, of FF-1 changes to 1. The output states (Q_2, Q_3 and Q_4) of other flip-flops remain at 0 as their inputs are at 0. The output of the register is 1000 and it is shown in Fig. 9.8.

Serial data input,
$A_3A_2A_1A_0 = 1101$

CLOCK	INPUT BIT	FF-1 Q_1	FF-2 Q_2	FF-3 Q_3	FF-4 Q_4	
———	CLEAR	0	0	0	0	
Trigger-1 ⤴	$A_0 = 1$	1	0	0	0	
Trigger-2 ⤴	$A_1 = 0$	0	1	0	0	Serial
Trigger-3 ⤴	$A_2 = 1$	1	0	1	0	data outputs
Trigger-4 ⤴	$A_3 = 1$	1	1	0	1	→ A_0
Trigger-5 ⤴	———	0	1	1	0	→ A_1
Trigger-6 ⤴	———	0	0	1	1	→ A_2
Trigger-7 ⤴	———	0	0	0	1	→ A_3

Fig. 9.8 Data shifting in Serial-in/Serial-out Shift register

Similarly, the next higher level bits (A_1, A_2 and A_3) are loaded to the input, D_1 of FF-1, bit by bit. The flip-flops are triggered after loading each bit. The changes in the output states of the flip-flops for Trigger-2, Trigger-3 and Trigger-4 are shown in Fig. 9.8. After Trigger-4, the bit at Q_4 of the FF-4 is the LSB (A_0) of input data and it is available for reading.

The shifting operations are continued with Trigger-5, Trigger-6 and Trigger-7. The higher level bits (A_1, A_2 and A_3) are available bit-by-bit serially at Q_4 of the FF-4 for reading. The clock signal is inhibited after Trigger-7.

9.3.4 Parallel-In/Parallel-Out Shift Register

The construction of Parallel-in/Parallel-out Shift registers is similar to Parallel-in/Serial-out Shift registers. The simplified schematic diagram of 4-bit Parallel-in/Parallel-out Shift register is shown in Fig. 9.9 and it is based on the data sheets of standard ICs (Ex.: 74LS395). The bits of input data are loaded to the flip-flops of the shift register simultaneously (parallel format). The data bits are available simultaneously (parallel format) at the output of the flip-flops of the shift register.

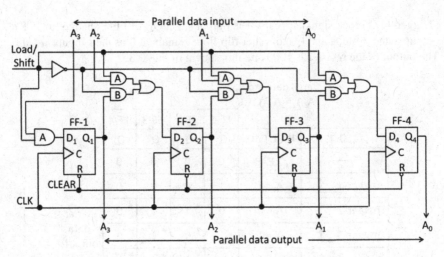

Fig. 9.9 Parallel-in/Parallel-out 4-bit Shift register

9.3.4.1 Operation

The operation of Parallel-in/Parallel-out Shift registers is similar to Parallel-in/Serial-out Shift registers. The output states of the flip-flops of 4-bit Parallel-in/Parallel-out Shift register are reset to 0 using Clear control before loading input data to the register. The Load/Shift control is set High for loading the four bit word, $A_3A_2A_1A_0$, 1101. The four AND gates, A, of the control gate circuit are activated and the four bits are loaded simultaneously to the flip-flops of the register. The MSB (A_3), 1, is loaded to D_1 of FF-1; the next lower bit (A_2), 1, is loaded to D_2 of FF-2; the next lower bit (A_1), 0, is loaded to D_3 of FF-3 and the LSB (A_0), 1, is loaded to D_4 of FF-4.

The Load/Shift control is set Low for shift operation. The three AND gates, B, of the control gate circuit are activated to facilitate the shifting of bits from one flip-flop to the next flip-flop. The flip-flops are triggered when the next positive edge (Trigger-1) of clock signal appears. The output states (Q_1, Q_2, Q_3 and Q_4), of the flip-flops are shown in Fig. 9.10. The bits at Q_1 to Q_4 of the flip-flops represent the parallel output data ($A_3A_2A_1A_0$) for reading. The clock signal is inhibited after Trigger-1.

		Parallel data input ↓	FF-1	FF-2	FF-3	FF-4
OPERATION	CLOCK	INPUT	Q_1	Q_2	Q_3	Q_4
CLEAR	——	——	0	0	0	0
LOAD Load/Shift=1	——	$D_1=A_3=1$ $D_2=A_2=1$ $D_3=A_1=0$ $D_4=A_0=1$	0	0	0	0
SHIFT Load/Shift=0	Trigger-1 ⎍↑	——	1	1	0	1

↓ A_3 ↓ A_2 ↓ A_1 ↓ A_0
Parallel data output

Fig. 9.10 Data shifting in Parallel-in/Parallel-out Shift register

9.4 Applications of Shift Registers

Shift registers are used in arithmetic circuits, data transmissions, counters and display drivers. Four applications of shift registers are presented. The applications are Ring counter, Johnson counter, Linear Feedback Shift Register (LFSR) and Serial adder.

9.4.1 Ring Counter

Ring counter generates code sequences that could be used for controlling digital circuits. It does not count the occurrences of events like binary counters. Ring counter is constructed by connecting the output of last flip-flop of shift register to the input of the first flip-flop of the shift register. The schematic diagram of 4-bit Ring counter is shown in Fig. 9.11a.

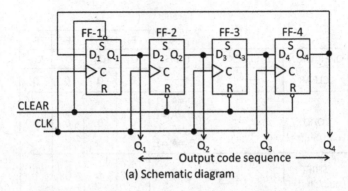

(a) Schematic diagram

OPERATION	FF-1	FF-2	FF-3	FF-4	OUTPUT CODE SEQUENCE
	Q_1	Q_2	Q_3	Q_4	$(Q_1Q_2Q_3Q_4)$
CLEAR	1	0	0	0	1000
Trigger-1 ⟋	0	1	0	0	0100
Trigger-2 ⟋	0	0	1	0	0010
Trigger-3 ⟋	0	0	0	1	0001

(b) Output code sequences

Fig. 9.11 4-bit Ring counter

The number of code sequences generated by Ring counter is equal to the number of bits handled by the shift register. For example, 4-bit shift register is capable of generating four code sequences. The generated code sequences are available as parallel output from the flip-flops.

9.4.1.1 Operation

The output, Q_4, of the flip-flop, FF-4, is connected to D_1 of FF-1. Clear control input is connected to the Reset inputs, R, of FF-2, FF-3 and FF-4 and to the Set input, S, of FF-1. When clear control signal is applied, Q_2, Q_3 and Q_4 of the flip-flops are reset to 0 and Q_1 of FF-1 is set to 1. The output code sequence is $Q_1Q_2Q_3Q_4$ i.e. 1000 and it is shown in Fig. 9.11b.

The flip-flops are triggered when the next positive edge (Trigger-1) of clock signal appears. The next output code sequence becomes 0100. The code sequence changes to 0010 and 0001 for Trigger-2 and Trigger-3 respectively. The output code sequences are shown in Fig. 9.11b. If Trigger-4 is applied to the flip-flops, the output code sequence becomes 1000. The circular loop from 1000 to 0001 continues for the subsequent set of clock triggers after Trigger-3. Ring counter is also called circular counter.

9.4.1.2 Modified Sequence Codes

The output code sequences of ring counter could be modified by setting more output states of flip-flops to 1. For example, the outputs (Q_1 and Q_2) of FF-1 and FF-2 could be set to 1 for generating the code sequences, 1100, 0110, 0011 and 1001. Similarly, the number of output code sequences of ring counter could be reduced. The outputs (Q_1 and Q_3) of FF-1 and FF-3 could be set to 1 for generating two code sequences, 1010 and 0101. More number of sequence codes could be generated by Johnson counter.

9.4.2 Johnson Counter

The schematic diagram of Johnson counter with 4-bit shift register is shown in Fig. 9.12. The construction of 4-bit Johnson counter are similar to Ring counter except that the inverted output (Q_4') of last flip-flop of shift register is connected to the input (D_1) of the first flip-flop, FF-1, of the shift register. Johnson counter with n-bit shift register generates 2n output code sequences. 4-bit Johnson counter generates eight output code sequences. Johnson counter is also called twisted ring counter or twisted tail counter.

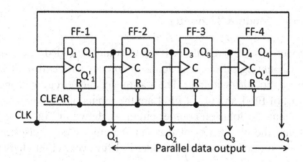

Fig. 9.12 4-bit Johnson counter

9.4.2.1 Operation

Clear input resets Q_1, Q_2, Q_3 and Q_4 of the flip-flops to 0. The output code sequence is $Q_1Q_2Q_3Q_4$ i.e. 0000 and it is shown in Fig. 9.13. As Q_4' is 1, the input, D_1, of FF-1 is also 1.

OPERATION	FF-1	FF-2	FF-3	FF-4	OUTPUT CODE SEQUENCE
	Q_1	Q_2	Q_3	Q_4	$(Q_1 Q_2 Q_3 Q_4)$
CLEAR	0	0	0	0	0000
Trigger-1 ⌐	1	0	0	0	1000
Trigger-2 ⌐	1	1	0	0	1100
Trigger-3 ⌐	1	1	1	0	1110
Trigger-4 ⌐	1	1	1	1	1111
Trigger-5 ⌐	0	1	1	1	0111
Trigger-6 ⌐	0	0	1	1	0011
Trigger-7 ⌐	0	0	0	1	0001

Fig. 9.13 Output code sequences of 4-bit Johnson counter

The flip-flops are triggered when the next positive edge (Trigger-1) of clock signal appears. The output code sequence becomes 1000. The code sequence changes for Trigger-2 to Trigger-7 are also shown in Fig. 9.13. If Trigger-8 is applied to the flip-flops, the output code sequence becomes 0000. The circular loop from 0000 to 0001 continues for subsequent set of clock triggers after Trigger-7.

9.4.2.2 Johnson Modulo-8 Counter

A simple decoder circuit with one AND gate could be added to Johnson counter in Fig. 9.12 for counting pulses. The Johnson counter with add-on decoder circuit becomes Johnson modulo-8 counter. The add-on decoder circuit has one AND gate. The outputs, Q_1', of FF-1 and Q_4' of FF-4 are the inputs of the AND gate. The add-on decoder without the Johnson counter circuit is shown in Fig. 9.14a. The inputs $(Q_1'$ and $Q_4')$ and the output (Z) of decoder for eight clock pulses are shown in Fig. 9.14b. The output of the add-on decoder is 1 for every eight clock pulses from 0 to 7.

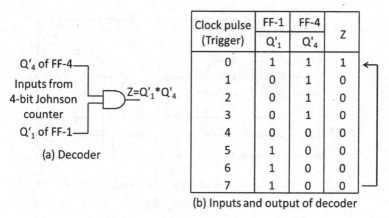

Clock pulse (Trigger)	FF-1 Q'_1	FF-4 Q'_4	Z
0	1	1	1
1	0	1	0
2	0	1	0
3	0	1	0
4	0	0	0
5	1	0	0
6	1	0	0
7	1	0	0

(a) Decoder

(b) Inputs and output of decoder

Fig. 9.14 Johnson 4-bit modulo-8 counter

9.4.3 Linear Feedback Shift Register

The basic construction of Linear Feedback Shift Register (LFSR) is similar to Ring counter or Johnson counter except for the feedback path between flip-flops. Feedback path in LFSR is designed using XOR gate and it generates more number of code sequences. n-bit LFSR generates ($2^n - 1$) output code sequences. 4-bit LFSR generates ($2^4 - 1$) i.e. 15 code sequences.

An example of feedback path with XOR gate in 4-bit LFSR circuit is shown in Fig. 9.15a. The general representation of the circuit is shown in Fig. 9.15b. Clear and clock signal lines are not shown in the general representation. The feedback gate junctions are called taps. One or more taps are used in LFSR for routing the outputs of the flip-flops to the inputs of flip-flops through XOR gate.

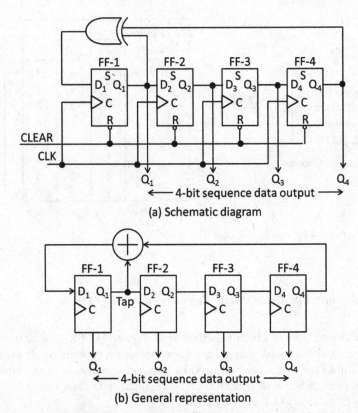

(a) Schematic diagram

(b) General representation

Fig. 9.15 Linear feedback 4-bit shift register

9.4.3.1 Operation

The 4-bit LFSR is reproduced in Fig. 9.16a. It is necessary to set at least one of the flip-flops to 1 for the operation of LFSR. Assume the output of FF-4 is set to 1. Clear input is given such that it resets Q_1, Q_2 and Q_3 of the flip-flops to 0 and Q_4 to 1. The output code sequence is $Q_1Q_2Q_3Q_4$ i.e. 0001 and it is shown in Fig. 9.16b. The outputs Q_4 of FF-4 and Q_1 of FF-1 are XORed, setting the input D_1 of FF-1 as 1 for next positive edge-triggering of clock signal.

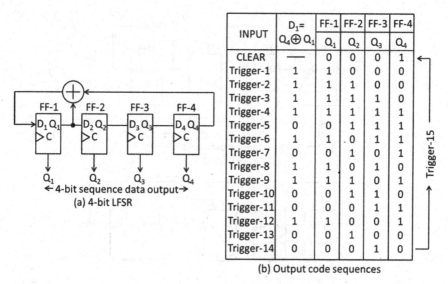

INPUT	$D_1=$ $Q_4 \oplus Q_1$	FF-1 Q_1	FF-2 Q_2	FF-3 Q_3	FF-4 Q_4
CLEAR	—	0	0	0	1
Trigger-1	1	1	0	0	0
Trigger-2	1	1	1	0	0
Trigger-3	1	1	1	1	0
Trigger-4	1	1	1	1	1
Trigger-5	0	0	1	1	1
Trigger-6	1	1	0	1	1
Trigger-7	0	0	1	0	1
Trigger-8	1	1	0	1	0
Trigger-9	1	1	1	0	1
Trigger-10	0	0	1	1	0
Trigger-11	0	0	0	1	1
Trigger-12	1	1	0	0	1
Trigger-13	0	0	1	0	0
Trigger-14	0	0	0	1	0

(b) Output code sequences

Fig. 9.16 Output code sequences of 4-bit LFSR

The flip-flops are triggered when the next positive edge (Trigger-1) of clock signal appears. The outputs of the flip-flops are shifted from left to right. The next output code sequence ($Q_1 Q_2 Q_3 Q_4$) becomes 1000. The process of XORing and triggering continues. The code sequences for Trigger-1 to Trigger-14 are shown in Fig. 9.16b. If Trigger-15 is applied to the flip-flops, the output code sequence becomes 0001 and the code sequences repeat for the subsequent set of clock triggers after Trigger-15.

It could be observed that the output code sequence, 0000, does not appear LFSR with XOR feedback path. XNOR gates instead of XOR could also be used in feedback paths. The output code sequence, 1111, does not appear LFSR with XNOR feedback path.

9.4.3.2 Types of Feedback Paths

Feedback paths are designed as per the functional requirements of LFSR. Two types of feedback paths with XOR gate are used. 4-bit LFSR with two types of feedback paths are shown in Fig. 9.17. The feedback path shown in 9.17a is external to LFSR and the feedback path in Fig. 9.17b is internal to LFSR.

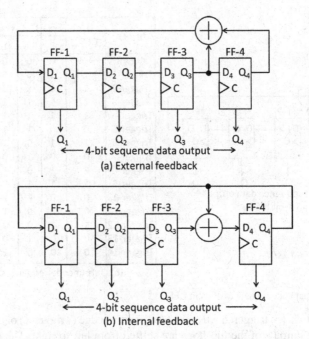

(a) External feedback

(b) Internal feedback

Fig. 9.17 Types of feedback path in LFSR

9.4.3.3   Linking Polynomial to LFSR

LFSR is used for cryptography as it is capable of generating pseudorandom code sequences. Higher level of encrypted data is realized in LFSR by designing with more bits (flip-flops) and taps (XOR feedback junctions) using XOR gates. The design is supported by linking polynomials to LFSR. The theory behind the linking of polynomial to LFSR and the associated theorems are explained in the reference [1].

An irreducible polynomial of degree, n, has $(2^n - 1)$ non-elementary sequences. The coefficients of polynomial decide the feedback taps for designing LFSR. For example, fourth order simple (primitive) polynomial represents 4-bit LFSR. The procedure for linking fourth order primitive polynomial to 4-bit LFSR is explained.

9.4.3.4   Fourth Order Polynomial with 4-Bit LFSR

The general form of fourth order polynomial is:

$$f(x) = x^4 + x^3 + x^2 + x^1 + x^0$$

The coefficients of the terms of the primitive polynomial are 1. Each term of the polynomial represents a connection to XOR gate between flip-flops. The output point of last flip-flop (FF-4) is represented by x^4 of the polynomial and the input point of first flip-flop (FF-1) is represented by x^0. The other terms represent feedback paths through XOR gates between flip-flops. The term, x^1, represents the feedback path between first and second (FF-1 and FF-2) flip-flops; x^2 represents the feedback path between second and third (FF-2 and FF-3) flip-flops and x^3 represents the feedback path between third and fourth (FF-3 and FF-4) flip-flops. If the coefficient of any term is 0, then is no feedback path to XOR gate between the adjacent flip-flops. For example, if the coefficient of x^2 is 0, there is no feedback path between second and third (FF-2 and FF-3) flip-flops. Three examples are provided linking fourth order polynomials to 4-bit LFSR.

9.4.3.5 Example-1

Let the fourth order primitive polynomial be:

$$f(x) = x^4 + x + 1$$

The polynomial is re-written as:

$$f(x) = 1 * x^4 + 0 * x^3 + 0 * x^2 + 1 * x^1 + 1 * x^0$$

The coefficient of x^1 is 1. A feedback path through XOR gate is required between the flip-flops, FF-1 and FF-2. The coefficients of x^3 and x^2 are 0. Hence, there are no feedback paths between other flip-flops. The schematic diagram of the LFSR for the polynomial is shown in Fig. 9.18.

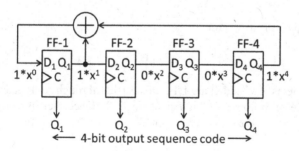

Fig. 9.18 Example-1: 4-bit LFSR

9.4.3.6 Example-2

Let the fourth order primitive polynomial be:

$$f(x) = x^4 + x^3 + 1$$

The polynomial is re-written as:

$$f(x) = 1 * x^4 + 1 * x^3 + 0 * x^2 + 0 * x^1 + 1 * x^0$$

The coefficient of x^3 is 1. A feedback path through XOR gate is required between the flip-flops, FF-3 and FF-4. The coefficients of x^2 and x^1 are 0. Hence, there are no feedback paths between other flip-flops. The schematic diagram of the LFSR for the polynomial is shown in Fig. 9.19.

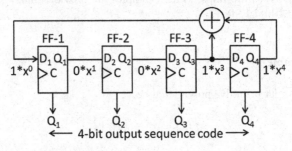

Fig. 9.19 Example-2: 4-bit LFSR

9.4.3.7 Example-3

Let the fourth order primitive polynomial be:

$$f(x) = x^4 + x^3 + x^1 + 1$$

The polynomial is re-written as:

$$f(x) = 1 * x^4 + 1 * x^3 + 0 * x^2 + 1 * x^1 + 1 * x^0$$

The coefficients of x^3 and x^1 are 1. Feedback paths through XOR gate are required between adjacent flip-flops. The coefficient of x^2 is 0. Hence, there is no feedback path between adjacent flip-flops. The schematic diagram of the LFSR for the polynomial is shown in Fig. 9.20.

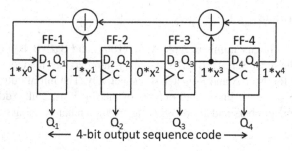

Fig. 9.20 Example-3: 4-bit LFSR

9.4.4 Serial Adder

A serial adder adds two binary numbers in pairs of bits from LSB to MSB of the numbers. The addition process is performed by two Serial-in/Serial-out shift registers (Register-A and Register-B), Full adder and carry bit flip-flop (FF-C). The simplified schematic diagram of 4-bit serial adder and shifting of the bits of the numbers in Full adder and Register-A are shown in Fig. 9.21.

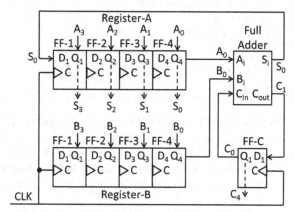

The final sum and carry bits are shown in
dashed arrows from Register-A and FF-C
$A_3A_2A_1A_0 + B_3B_2B_1B_0 = C_4S_3S_2S_1S_0$

Fig. 9.21 4-bit Serial Adder

Initially, the binary numbers are loaded to the outputs of flip-flops in Register-A and Register-B respectively. The LSBs, A_0 and B_0 are the inputs to Full adder (FA). The initial input carry bit, C_0 is 0. The FA adds the LSBs and C_0. The outputs of the FA are S_1 and C_1. S_1 is fed back to D_1 of FF-1 of Register-A. The above are initial status of the adder before clock triggering and they are shown in Fig. 9.21. The operation of serial adder is explained for adding two 4-bit binary numbers.

9.4.4.1 Operation

Assume the 4-bit binary numbers $A_3A_2A_1A_0$ (1001) and $B_3B_2B_1B_0$ (1101) need to be added by the 4-bit Serial Adder. The binary numbers are loaded to the outputs of flip-flops in Register-A and Register-B respectively. The initial output of FF-C is set to 0 i.e. C_0 is 0. The inputs and outputs of the flip-flops, Full Adder and FF-C after loading the numbers and after triggering are shown below using state transition analysis.

9.4.4.2 Loading Binary Numbers

The inputs and outputs of the flip-flops, Full Adder and FF-C are:

Flip-flops of Register-A:
$Q_1 = A_3 = 1; Q_2 = A_2 = 0; Q_3 = A_1 = 0; Q_4 = A_0 = 1$
$D_1 = S_0 \text{ (FA)} = 0; D_2 = Q_1 = 1; D_3 = Q_2 = 0; D_4 = Q_3 = 0$
Flip-flops of Register-B:
$Q_1 = B_3 = 1; Q_2 = B_2 = 1; Q_3 = B_1 = 0; Q_4 = B_0 = 1$
$D_1 = X \text{ (don't care)}; D_2 = Q_1 = 1; D_3 = Q_2 = 1; D_4 = Q_3 = 0$
Full Adder:
$A_i = Q_4 = 1; B_i = Q_4 = 1; C_{in} \text{ (Q1 of FF-C)} = 0$
$S_i = S_0 = 1 + 1 + 0 = 0; C_{out} = C_1 = 1$
FF-C: $D_1 = C_1 = 1; Q_1 = C_0 = 0$

9.4.4.3 Trigger-1

When the next positive edge (Trigger-1) of clock signal appears, the flip-flops of Registers (A and B) and the carry bit flip-flop (FF-C) are triggered. The bits at the inputs of the flip-flops of the Registers and FF-C are right-shifted. The inputs to FA are A_1, B_1 and C_0. The outputs of the FA are S_1 and C_1. S_1 is fed back to D_1 of FF-1 of Register-A. The inputs and outputs of the flip-flops, Full Adder and FF-C after Trigger-1 are:

Flip-flops of Register-A:
$Q_1 = S_0 = 0; Q_2 = 1; Q_3 = 0; Q_4 = 0$
$D_1 = S_1 \text{ (FA)} = 1; D_2 = Q_1 = S_0 = 0; D_3 = Q_2 = 1; D_4 = Q_3 = 0$
Flip-flops of Register-B:
$Q_1 = X; Q_2 = 1; Q_3 = 1; Q_4 = 0$
$D_1 = X; D_2 = Q_1 = X; D_3 = Q_2 = 1; D_4 = Q_3 = 1$
Full Adder:
$A_i = Q_4 = 0; B_i = Q_4 = 0; C_{in} \text{ (Q1 of FF-C)} = 1$
$S_i = S_1 = 0 + 0 + 1 = 1; C_{out} = C_2 = 0$
FF-C: $D_1 = C_2 = 0; Q_1 = C_1 = 1$

9.4.4.4 Trigger-2

When Trigger-2 appears, the bits at the inputs of the flip-flops of the Registers and FF-C are right-shifted. The outputs of the FA are S_2 and C_2. S_2 is fed back to D_1 of FF-1 of Register-A. The inputs and outputs of the flip-flops, Full Adder and FF-C after Trigger-2 are:

Flip-flops of Register-A:
$Q_1 = S_1 = 1; Q_2 = S_0 = 0; Q_3 = 1; Q_4 = 0$
$D_1 = S_2 \text{ (FA)} = 1; D_2 = Q_1 = S_1 = 1; D_3 = S_0 = Q_2 = 0; D_4 = Q_3 = 1$
Flip-flops of Register-B:
$Q_1 = X; Q_2 = X; Q_3 = 1; Q_4 = 1$
$D_1 = X; D_2 = Q_1 = X; D_3 = Q_2 = X; D_4 = Q_3 = 1$
Full Adder:
$A_i = Q_4 = 0; B_i = Q_4 = 1; C_{in} \text{ (Q1 of FF-C)} = 0$
$S_i = S_2 = 0 + 1 + 0 = 1; C_{out} = C_3 = 0$
FF-C: $D_1 = C_3 = 0; Q_1 = C_2 = 0$

9.4.4.5 Trigger-3

When Trigger-3 appears, the bits at the inputs of the flip-flops of the Registers and FF-C are right-shifted. The outputs of the FA are S_3 and C_3. S_3 is fed back to D_1 of FF-1 of Register-A. The inputs and outputs of the flip-flops, Full Adder and FF-C after Trigger-3 are:

Flip-flops of Register-A:
$Q_1 = S_2 = 1; Q_2 = S_1 = 1; Q_3 = S_0 = 0; Q_4 = 1$
$D_1 = S_3 \text{ (FA)} = 0; D_2 = Q_1 = S_2 = 1; D_3 = Q_2 = S_1 = 1; D_4 = Q_3 = S_0 = 0$
Flip-flops of Register-B:
$Q_1 = X; Q_2 = X; Q_3 = X; Q_4 = 1$
$D_1 = X; D_2 = Q_1 = X; D_3 = Q_2 = X; D_4 = Q_3 = X$
Full Adder:
$A_i = Q_4 = 1; B_i = Q_4 = 1; C_{in} \text{ (Q1 of FF-C)} = 0$
$S_i = S_3 = 1 + 1 + 0 = 0; C_{out} = C_4 = 1$
FF-C: $D_1 = C_4 = 1; Q_1 = C_3 = 0$

9.4.4.6 Trigger-4

When Trigger-4 appears, the bits at the inputs of the flip-flops of the Registers and FF-C are right-shifted. The outputs of the FA are S_4 and C_4. S_4 is fed back to D_1 of FF-1 of Register-A. The inputs and outputs of the flip-flops, Full Adder and FF-C after Trigger-4 are:

Flip-flops of Register-A:

$Q_1 = S_3 = 0; Q_2 = S_2 = 1; Q_3 = S_1 = 1; Q_4 = S_0 = 0$

$D_1 = S_3 (FA) = 0; D_2 = Q_1 = S_2 = 0; D_3 = Q_2 = S_1 = 1; D_4 = Q_3 = S_0 = 0$

Flip-flops of Register-B:

$Q_1 = X; Q_2 = X; Q_3 = X; Q_4 = X$

$D_1 = X; D_2 = Q_1 = X; D_3 = Q_2 = X; D_4 = Q_3 = X$

Full Adder:

$A_i = Q_4 = 0; B_i = Q_4 = X; C_{in} (Q1 \text{ of } FF\text{-}C) = 1$

$S_i = S_4 = 1 + X + 0 = X; C_{out} = C_5 = X$

FF-C: $D_1 = C_5 = X; Q_1 = C_4 = 1$

The clock signal is inhibited after Trigger-4. The summary of the state transition analysis for Full Adder and Register-A is shown in Fig. 9.22. The final result of adding the two binary numbers is available after Trigger-4 at the outputs of FF-C (C_4) and Register-A ($Q_1Q_2Q_3Q_4$). The outputs are shown with dashed arrows in 9.21.

$A_3A_2A_1A_0 = 1001$ $B_3B_2B_1B_0 = 1101$

INPUT	Full adder				Register-A			
	A_i	B_i	C_{in}	S_i	Q_1	Q_2	Q_3	Q_4
LOAD	A_0 (1)	B_0 (1)	C_0 (0)	S_0 (0)	A_3 (1)	A_2 (0)	A_1 (0)	A_0 (1)
Trigger-1	A_1 (0)	B_1 (0)	C_1 (1)	S_1 (1)	S_0 (0)	A_3 (1)	A_2 (0)	A_1 (0)
Trigger-2	A_2 (0)	B_2 (1)	C_2 (0)	S_2 (1)	S_1 (1)	S_0 (0)	A_3 (1)	A_2 (0)
Trigger-3	A_3 (1)	B_3 (1)	C_3 (0)	S_3 (0)	S_2 (1)	S_1 (1)	S_0 (0)	A_3 (1)
Trigger-4	0	X	C_4 (1)	X	S_3 (0)	S_2 (1)	S_1 (1)	S_0 (0)

The bit values are shown in brackets.

For example, $A_0 = 1$ is shown as A_0 (1)

X: Don't care

$A_3A_2A_1A_0 + B_3B_2B_1B_0 = C_4S_3S_2S_1S_0 = 10110$

Fig. 9.22 Shifting of bits in 4-bit Serial Adder

$A_3A_2A_1A_0 + B_3B_2B_1B_0 = C_4Q_1Q_2Q_3Q_4 = C_4S_3S_2S_1S_0$

$1001 + 1101 = 10110$

9.5   Universal Shift Register

The basic types of shift registers are unidirectional shift registers. The input data is shifted from left to right by clock triggers in the flip-flops of shift registers. Applications such as code conversion and arithmetic units require bidirectional shift registers for manipulating the bits of data to obtain desired results. For example, left-shifting the bits of binary number by one bit position is equivalent to multiplication by 2 and right-shifting the bits of data by one bit position is equivalent to dividing by 2.

Universal shift register is a bidirectional shift register and it is used for many applications. It has many additional features. The features of standard ICs (Ex.: 74LS194) for Universal shift register are:

- Hold (Do nothing) mode
- Shift right (Serial-in/Serial-out, Serial-in/Parallel-out and Parallel-in/Serial-out)
- Shift left (Serial-in/Serial-out, Serial-in/Parallel-out and Parallel-in/Serial-out)
- Parallel-in/Parallel-out

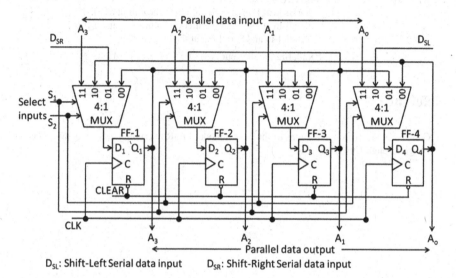

Fig. 9.23 4-bit Universal shift register

Data transfers are performed synchronously with clock signal in Universal shift register. The simplified schematic diagram of 4-bit Universal shift register is shown in Fig. 9.23. The schematic diagram is based on the datasheets of ICs for universal shift registers.

9.5.1 Mode Select Function

The basic functioning of 4-bit Universal shift register is similar to basic shift registers except that the serial and parallel data inputs are routed through 4:1 MUX. The required mode of operation such as Serial-in/Parallel-out with left-shift or Serial-in/Serial-out with right-shift is selected by the select inputs to the MUX. The controls and the operation of Universal shift register are shown in the mode select function in Table 9.1. The shifting operations are controlled by clock triggers as explained in the operation of basic shift registers.

Table 9.1 Mode select inputs and operations in Universal shift register

Select input	Operation
$S_1S_2 = 00$	Hold (No change in the output states of flip-flops)
$S_1S_2 = 01$	Shift right: – Serial-in at D_{SR} and Serial-out at the output of FF-4 – Serial-in at D_{SR} and Parallel-out at the outputs of FFs – Parallel-in and Serial-out at the output of FF-4
$S_1S_2 = 10$	Shift left: – Serial-in at D_{SL} and Serial-out at the output of FF-1 – Serial-in at D_{SL} and Parallel-out at the outputs of FFs – Parallel-in and Serial-out at the output of FF-1
$S_1S_2 = 11$	Parallel-in/Parallel-out

Reference

1. Gardner DG (2016) Applications of the Galois model LFSR in cryptography, Doctoral Thesis, Loughborough University, UK

Chapter 10
Counters

Abstract Two or more flip-flops are cascaded in counters. n-bit counter requires n flip-flops. The output states of the flip-flops of counters change for a pre-determined number of count sequences. The operation of asynchronous and synchronous counters is explained. The operation of up-down counters and decoders for counters is also explained. Cascading counters for realizing higher modulus counters is explained and illustrated with examples. The application of cascading counters for digital clock is presented. The design of Mod-6 binary counter and Mod-6 Gray code counter is illustrated with examples.

10.1 Introduction

Counters are used for applications, such as timers, elapsed time measurements and frequency counters. Elapsed time in terms of clock periods is measured for executing computer instructions with reference to clock signal. Counters are used for totaling the number of pulses to measure frequency and the test equipment is popularly known as frequency counter.

The basic construction of counter is similar to register. Flip-flops are cascaded in counters. n-bit counter requires n flip-flops. Unlike registers, the output states of the flip-flops of counters change for a pre-determined number of count sequences. Changes in the states of the flip-flops are caused by triggers. The triggers could be from clock signal or from the outputs of previous flip-flops.

10.1.1 Understanding Counters

The output states of cascaded flip-flops in parallel format represent the output of counter. Assume that the bits of the flip-flops are right-shifted. The output state of first flip-flop is the LSB of the output data of counter and that of the last flip-flop

D. Natarajan, *Fundamentals of Digital Electronics*,
Lecture Notes in Electrical Engineering 623,
https://doi.org/10.1007/978-3-030-36196-9_10

is MSB. Generally, each clock trigger increments the binary output data of counter by one bit. For example, consider a 4-bit BCD counter. Initially, the output data of the counter is reset to 0000. The output becomes 0001 after first trigger; the output becomes 0010 after second trigger; the output becomes 0011 after third trigger and so on. Each increment is equivalent to counting the number of clock pulses applied to counter. The count sequences of 4-bit BCD counter are from 0000 to 1001 i.e. 0 to 9. Counting could be demonstrated by connecting the output of flip-flops of the counter to 7-segment LED display through a BCD to 7-segment decoder/driver (IC 7447 or 7448). Section 4.4.6.3 could be referred for the details of connections. The 7-segment display shows the decimal equivalent counting from 0 to 9. 1 Hz clock signal could be used for triggering to view the display comfortably.

10.1.1.1   Modulus of Counter

The modulus of a counter is the number of states available as outputs from the counter. In general, the maximum number of states that could be generated by n-bit counter is 2^n and the modulus of the counter is 2^n. For example, consider 4-bit counter. Assuming that the counter generates the maximum number of states i.e. 16, the counter is termed as Mod-16 counter.

10.1.1.2   Classification of Counters

Counters are classified as asynchronous and synchronous. The classification is based on the method of triggering the flip-flops of counters. In asynchronous counters, the first flip-flop is triggered by clock signal and the subsequent flip-flops are triggered by the outputs of previous flip-flops. In synchronous counters, all the flip-flops are triggered simultaneously by common clock signal. The datasheets of ICs for counters present logic diagrams with JK or D or T flip-flops. The operation of asynchronous and synchronous counters is presented with T flip-flops.

10.2   Asynchronous Counters

The operation of two types of asynchronous counters, namely, 4-bit Binary ripple counter and 4-bit BCD ripple counter with T flip-flops are presented. The toggling states of T flip-flop are used in the counters. The output of the flip-flop toggles with successive clock triggers when the input (T) is 1.

10.2.1 Binary Ripple Counter

The schematic diagram of simplified 4-bit binary ripple counter using four T flip-flops is shown in Fig. 10.1a. The flip-flops are negative edge-triggered. The T inputs of the flip-flops are connected to V_{CC} for obtaining toggling outputs with triggers. The flip-flop, FF-1 is triggered by clock signal. FF-2 is triggered by the output of FF-1. FF-3 is triggered by the output of FF-2. FF-4 is triggered by the output of FF-3. The sequential triggering is equated to ripples and hence the counter is called binary ripple counter. The output of the counter is $Q_3Q_2Q_1Q_0$ and it is shown in the figure.

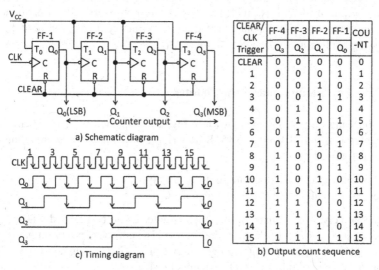

CLEAR/ CLK Trigger	FF-4 Q_3	FF-3 Q_2	FF-2 Q_1	FF-1 Q_0	COU -NT
CLEAR	0	0	0	0	0
1	0	0	0	1	1
2	0	0	1	0	2
3	0	0	1	1	3
4	0	1	0	0	4
5	0	1	0	1	5
6	0	1	1	0	6
7	0	1	1	1	7
8	1	0	0	0	8
9	1	0	0	1	9
10	1	0	1	0	10
11	1	0	1	1	11
12	1	1	0	0	12
13	1	1	0	1	13
14	1	1	1	0	14
15	1	1	1	1	15

a) Schematic diagram

c) Timing diagram

b) Output count sequence

Fig. 10.1 Asynchronous 4-bit Binary ripple counter

10.2.1.1 Operation

Initially, the output states of the flip-flops of the 4-bit Binary ripple counter are reset to 0 using Clear control. The output state of FF-1 changes from 0 to 1 with trigger-1. The output state of FF-2 remains at 0 as the trigger from FF-1 is positive. The output states of other flip-flops also remain at 0. The output of the counter is $Q_3Q_3Q_1Q_0$ i.e. 0001.

The output state of FF-1 changes from 1 to 0 with clock Trigger-2. It provides negative edge-triggering to FF-2. The output state of FF-2 changes to 1. The output states of other flip-flops continue to remain at 0. The output of the counter is 0010. Similarly, the toggling outputs occur in the output states of the flip-flops for the clock triggers from 3 to 15. The binary count sequences and the equivalent decimal counts are shown in Fig. 10.1b. The 4-bit Binary ripple counter has 16 states. The modulus of the counter is 16.

The timing diagram of 4-bit Binary ripple counter is shown in Fig. 10.1c. The diagram shows the toggling output states of the flip-flops. When clock Trigger-16 is applied to the FF-1 of the counter, the outputs of the flip-flops are reset to 0. Propagation delay is ignored in the timing diagram.

10.2.1.2   Frequency Divider

It could be observed in the timing diagram, Fig. 10.1c, of 4-bit counter that clock signal frequency at the output of FF-1 is divided by a factor of 2; it is divided by 4 at the output of FF-2; it is divided by 8 at the output of FF-3 and it is divided by 16 at the output of FF-4. Counters are used as frequency dividers in digital signal processing applications.

10.2.1.3   Up-Counter and Down-Counter

The 4-bit Binary ripple counter shown in Fig. 10.1 is called Mod-16 Binary ripple up-counter as it counts from 0 to 15. Binary ripple down-counter for counting from 15 to 0 is realized by:

- The outputs of the flip-flops are set to 1 initially
- The flip-flops are positive edge-triggered.

Alternatively, 4-bit Binary ripple down-counter could be realized with negative edge-triggering for the flip-flops of the counter. The complementary output of FF-1 (Q_0') triggers FF-2; the complementary output of FF-2 (Q_1') triggers FF-3 and the complementary output of FF-3 (Q_2') triggers FF-4.

10.2.2   BCD Ripple Counter

The basic construction of 4-bit BCD is same as that of 4-bit Binary ripple counter. The BCD Counter counts from 0000 to 1001. The modulus of BCD ripple counter is 10. The schematic diagram of simplified 4-bit BCD ripple counter using four T flip-flops is shown in Fig. 10.2a. The flip-flops are negative edge-triggered. The T inputs of the flip-flops are connected to V_{CC} for obtaining toggling outputs with triggers. The output of the counter is $Q_3Q_2Q_1Q_0$. The counter is also called Decade ripple counter.

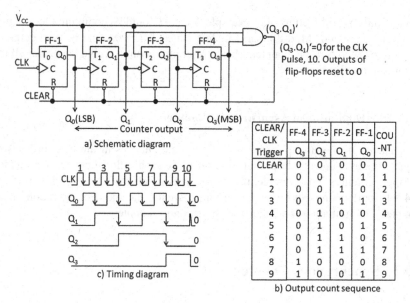

Fig. 10.2 Asynchronous 4-bit BCD ripple counter

The outputs (Q_1 and Q_3) of FF-2 and FF-4 are connected to NAND gate. Although 4-bit counter is capable of generating 16 output states, the available number of output states is limited to ten in BCD counter by the NAND gate.

10.2.2.1 Operation

Initially, the output states of the flip-flops of the 4-bit BCD ripple counter are reset to 0 using Clear control. The operation of BCD ripple counter is same as that of Binary ripple counter from Clear to clock trigger-9. The binary count sequences from 0000 to 1001 and the equivalent decimal counts are shown in Fig. 10.2b.

When clock trigger-10 is applied to FF-1, the output of the counter becomes 1010. The output states of Q_3 and Q_1 are 1. Q_3 and Q_1 drive the NAND gate. The NAND gate outputs active Low signal. The active Low signal resets the output states of the flip-flops of the BCD ripple counter to 0.

The timing diagram of the BCD ripple counter is shown in Fig. 10.2c. It could be observed in the figure that the output of Q1 goes to 1 momentarily before it is reset to 0 by the active Low signal from the NAND gate.

10.2.2.2 Standard ICs

Standard ICs are available for binary and BCD ripple counters. 74LS93 is 4-bit Binary ripple counter and 74LS90 is 4-bit BCD ripple counter. The datasheets of the ICs could be referred for logic and pin connection diagrams of the counters. The devices have added features. For example, 74LS93 could be configured to function either as Mod-16 or Mod-8 binary ripple counter. 74LS90 could be configured as Mod-10 or Mod-5 ripple counter. Master reset provision is available to clear all the outputs of the flip-flops.

10.2.2.3   Limitation of Asynchronous Counters

Propagation delays are specified for various state changes in the IC datasheets of asynchronous counters. The propagation delay of each flip-flop is cumulatively added in ripple counters. For example, if the propagation delay of each flip-flop is 15 ns, a 4-bit ripple counter has a total of 60 ns, reducing the speed of operation of the counter. The limitation is overcome in synchronous counters.

10.3   Synchronous Counters

Like asynchronous counters, synchronous counters also use flip-flops in cascaded form. The flip-flops of synchronous counters are edge-triggered simultaneously by clock signal. The operation of two types of synchronous counters, namely, 4-bit Binary counter and 4-bit BCD counter are presented.

10.3.1   Binary Counter

The schematic diagram of simplified synchronous 4-bit binary counter using four T flip-flops is shown in Fig. 10.3. Simultaneous triggering of the flip-flops using common clock signal is also shown in the figure. The flip-flops of the counter are negative edge-triggered.

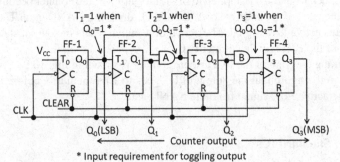

Fig. 10.3 Synchronous 4-bit Binary counter

T_0 of FF-1 is connected to V_{CC} for the operation of synchronous Binary counter; otherwise, the output of the counter would remain at 0000. Q_0 of FF-1 is connected to T_1 of FF-2. Q_0 and Q_1 are connected to T_2 of FF-2 through AND gate, A. The output of the AND gate, A, and Q_2 are connected to T_3 of FF-3 through another AND gate, B. The outputs of the flip-flops toggle when the input logic states of the flip-flops are at 1 before the arrival of clock trigger. The input logic states of the flip-flops are at 1 when:

T_0 remains at 1 as it is connected V_{CC}.
$T_1 = 1$ when $Q_0 = 1$
$T_2 = 1$ when $Q_0 Q_1 = 1$
$T_3 = 1$ when $Q_0 Q_1 Q_2 = 1$.

The above conditions are shown in Fig. 10.3. The output of the counter is $Q_3 Q_2 Q_1 Q_0$. The operation of the counter is explained using state transition analysis.

10.3.1.1 State Transition Analysis

Initially, the output states of the flip-flops of the synchronous 4-bit Binary counter are reset to 0 using Clear control. Hence, the logic status of T_1, T_2 and T_3 is 0. The logic status of T_0 is 1 as it is connected to V_{CC}. When clock trigger-1 appears:

– Q_0 changes to 1
– Q_1, Q_2 and Q_3 remains at 0
– $Q_3 Q_2 Q_1 Q_0 = 0001$
– $T_0 = 1$; $T_1 = Q_0 = 1$; $T_2 = Q_0 Q_1 = 0$; $T_3 = Q_0 Q_1 Q_2 = 0$

   Clock trigger-2:

– Q_0 changes to 0; Q_1 changes to 1
– Q_2 and Q_3 remains at 0
– $Q_3 Q_2 Q_1 Q_0$, is 0010
– $T_0 = 1$; $T_1 = Q_0 = 0$; $T_2 = Q_0 Q_1 = 0$; $T_3 = Q_0 Q_1 Q_2 = 0$

   Clock trigger-3:

– Q_0 changes to 1; Q_1 remains at 1
– Q_2 and Q_3 remains at 0
– $Q_3 Q_2 Q_1 Q_0$, is 0011
– $T_0 = 1$; $T_1 = Q_0 = 1$; $T_2 = Q_0 Q_1 = 1$; $T_3 = Q_0 Q_1 Q_2 = 0$.

Similar analysis could be done for clock triggers from 3 to 15. Q_0 toggles for each clock trigger. Q_1 toggles for the triggers 2, 4, 6, 8, 10, 12 and 14 as $T_1 = Q_0 = 1$ for the triggers. Q_2 toggles for the triggers 4, 8 and 12 as $T_2 = Q_0 . Q_1 = 1$ for the triggers. Q_3 toggles for the trigger-8 as $T_3 = Q_0 Q_1 Q_2 = 1$ for the trigger. Clock trigger-16 resets the counter and the output of the counter becomes 0000. Timing diagram is shown in Fig. 10.4a. The binary count sequences and the equivalent decimal counts are shown in Fig. 10.4b.

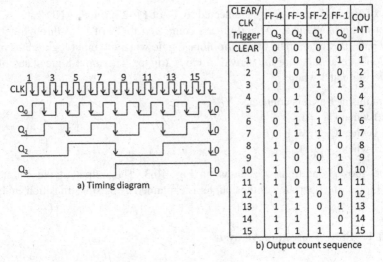

CLEAR/ CLK	FF-4	FF-3	FF-2	FF-1	COU -NT
Trigger	Q_3	Q_2	Q_1	Q_0	
CLEAR	0	0	0	0	0
1	0	0	0	1	1
2	0	0	1	0	2
3	0	0	1	1	3
4	0	1	0	0	4
5	0	1	0	1	5
6	0	1	1	0	6
7	0	1	1	1	7
8	1	0	0	0	8
9	1	0	0	1	9
10	1	0	1	0	10
11	1	0	1	1	11
12	1	1	0	0	12
13	1	1	0	1	13
14	1	1	1	0	14
15	1	1	1	1	15

a) Timing diagram

b) Output count sequence

Fig. 10.4 Output states of synchronous 4-bit Binary counter

10.3.2 BCD Counter

The schematic diagram of simplified synchronous 4-bit BCD counter using four T flip-flops is shown in Fig. 10.5. Simultaneous triggering of the flip-flops using common clock signal is also shown in the figure. The flip-flops of the counter are positive edge-triggered. The BCD counter is also called Decade counter. T_0 of FF-1 is connected to V_{CC}. The output of each flip-flop is connected to the input of next flip-flop through an AND gate circuit. The complemented output, Q'_3, of FF-3 is connected to the AND gate, A, for limiting the number of output states to ten (0000 to 1001). The outputs of the flip-flops toggle when the input logic states of the flip-flops are at 1 before the arrival of clock trigger. The input logic states of the flip-flops are at 1 when:

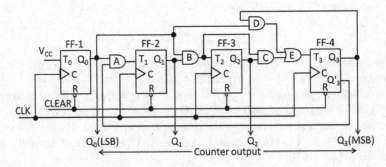

Fig. 10.5 Synchronous 4-bit BCD counter

T_0 remains at 1 as it is connected V_{CC}.
$T_1 = 1$ when $Q_0 = 1$
$T_2 = 1$ when $Q_0Q_1 = 1$
$T_3 = 1$ when $Q_0Q_1Q_2 = 1$.

The output of the counter is $Q_3Q_2Q_1Q_0$. The operation of the counter is explained using state transition analysis.

10.3.2.1 State Transition Analysis

Initially, the output states of the flip-flops of the synchronous 4-bit Binary counter are reset to 0 using Clear control. Hence, the logic status of T_1, T_2 and T_3 is 0. The logic status of T_0 is 1 as it is connected to V_{CC}. When clock trigger-1 appears:

- Q_0 changes to 1
- Q_1, Q_2 and Q_3 remains at 0; $Q'_3 = 1$
- $Q_3Q_2Q_1Q_0 = 0001$
- $T_0 = 1$; $T_1 = Q_0Q'_3 = 1$; $T_2 = Q_0Q_1 = 0$; $T_3 = (Q_0Q_3 + Q_0Q_1Q_2) = 0$

Clock trigger-2:

- Q_0 changes to 0; Q_1 changes to 1
- Q_2 and Q_3 remains at 0; $Q'_3 = 1$
- $Q_3Q_2Q_1Q_0 = 0010$
- $T_0 = 1$; $T_1 = Q_0Q'_3 = 0$; $T_2 = Q_0Q_1 = 0$; $T_3 = (Q_0Q_3 + Q_0Q_1Q_2) = 0$

Clock trigger-3:

- Q_0 changes to 1
- Q_1 remains at 1; Q_2 and Q_3 remains at 0; $Q'_3 = 1$
- $Q_3Q_2Q_1Q_0 = 0011$
- $T_0 = 1$; $T_1 = Q_0Q'_3 = 1$; $T_2 = Q_0Q_1 = 1$; $T_3 = (Q_0Q_3 + Q_0Q_1Q_2) = 0$

Clock trigger-4:

- Q_0 and Q_1 change to 0; Q_2 changes to 1
- Q_3 remains at 0; $Q'_3 = 1$
- $Q_3Q_2Q_1Q_0 = 0100$
- $T_0 = 1$; $T_1 = Q_0Q'_3 = 0$; $T_2 = Q_0Q_1 = 0$; $T_3 = (Q_0Q_3 + Q_0Q_1Q_2) = 0$

Clock trigger-5:

- Q_0 changes to 1
- Q_1 remains at 0; Q_2 remains at 1; and Q_3 remains at 0; $Q'_3 = 1$
- $Q_3Q_2Q_1Q_0 = 0101$
- $T_0 = 1$; $T_1 = Q_0Q'_3 = 1$; $T_2 = Q_0Q_1 = 0$; $T_3 = (Q_0Q_3 + Q_0Q_1Q_2) = 0$

Clock trigger-6:

- Q_0 changes to 0; Q_1 changes to 1

- Q_2 remains at 1; Q_3 remains at 0; $Q_3' = 1$
- $Q_3Q_2Q_1Q_0 = 0110$
- $T_0 = 1; T_1 = Q_0Q_3' = 0; T_2 = Q_0Q_1 = 0; T_3 = (Q_0Q_3 + Q_0Q_1Q_2) = 0$

Clock trigger-7:

- Q_0 changes to 1
- Q_1 and Q_2 remains at 1; Q_3 remains at 0; $Q_3' = 1$
- $Q_3Q_2Q_1Q_0 = 0111$
- $T_0 = 1; T_1 = Q_0Q_3' = 1; T_2 = Q_0Q_1 = 1; T_3 = (Q_0Q_3 + Q_0Q_1Q_2) = 1$

Clock trigger-8:

- Q_0, Q_1 and Q_2 changes to 0; Q_3 changes to 1; $Q_3' = 1$
- $Q_3Q_2Q_1Q_0 = 1000$
- $T_0 = 1; T_1 = Q_0Q_3' = 0; T_2 = Q_0Q_1 = 0; T_3 = (Q_0Q_3 + Q_0Q_1Q_2) = 0$

Clock trigger-9:

- Q_0 changes to 1
- Q_1 and Q_2 remains at 0; Q_3 remains at 1; $Q_3' = 0$
- $Q_3Q_2Q_1Q_0 = 1001$
- $T_0 = 1; T_1 = Q_0Q_3' = 0; T_2 = Q_0Q_1 = 0; T_3 = (Q_0Q_3 + Q_0Q_1Q_2) = 1$

Clock trigger-10:

- Q_0 and Q_3 changes to 0; $Q_3' = 1$
- Q_1 and Q_2 remains at 0
- $Q_3Q_2Q_1Q_0 = 0000$.

The clock trigger-10 resets the counter and $Q_3Q_2Q_1Q_0$ becomes 0000. Timing and state transition diagrams are shown in Fig. 10.6. The state transition diagram shows the output count sequence for clock triggers.

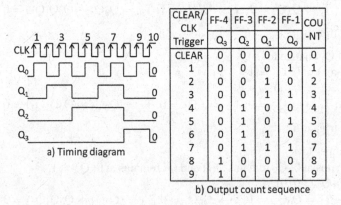

CLEAR/ CLK Trigger	FF-4 Q_3	FF-3 Q_2	FF-2 Q_1	FF-1 Q_0	COU -NT
CLEAR	0	0	0	0	0
1	0	0	0	1	1
2	0	0	1	0	2
3	0	0	1	1	3
4	0	1	0	0	4
5	0	1	0	1	5
6	0	1	1	0	6
7	0	1	1	1	7
8	1	0	0	0	8
9	1	0	0	1	9

a) Timing diagram

b) Output count sequence

Fig. 10.6 Output states of synchronous 4-bit BCD counter

10.3.2.2 Standard ICs

Standard ICs are available for synchronous binary and BCD counters in TTL and CMOS technologies. 74LS163 is 4-bit Binary counter and 74LS162 is 4-bit BCD counter. The ICs have added features. Control inputs are available for enabling and inhibiting count mode. Any count state could be retained with the control inputs.

10.3.3 Up-Down Counters

The synchronous 4-bit Binary and BCD counters are explained in the previous sections are up-counters. The count sequence is from 0000 to 1111 or from 0000 to 1001 as applicable to the counters. The schematic diagrams of the counters could be modified by adding additional gate circuits for realizing synchronous up-down binary or BCD counters. Up-down binary counter is capable of counting from 0000 to 1111 or vice versa. Similarly, up-down BCD counter is capable of counting from 0000 to 1001 or vice versa.

10.3.3.1 Standard ICs

Standard ICs are available for synchronous up/down binary and BCD counters. 74LS190 is a 4-bit up/down BCD counter and 74LS191 is a 4-bit up/down Binary counter. The flip-flops of the ICs are negative edge-triggered. The ICs have count-enable input. If Count-enable input is set High, counting is inhibited. The direction of the count is determined by the level of Up/Down input. When Up/Down input is set Low, the counter functions as up-counter. When Up/Down input is set High, the counter functions as down-counter. The datasheets of the ICs could be referred for additional information.

10.3.3.2 Schematic Diagram of up-Down Counter

The schematic diagram of synchronous up-down binary counter is shown in Fig. 10.7. The flip-flops are positive edge-triggered. Up-Down count sequence is controlled by the input, U/D'. When U/D' is set High, up-counting is realized. When U/D' is set Low, down-counting is realized.

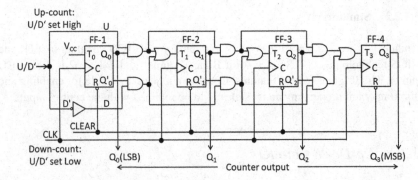

Fig. 10.7 Synchronous 4-bit Binary up-down counter

10.3.3.3   Operation of up-Down Counter

The output states of flip-flops change with clock trigger only when their input logic states are at 1. The input logic states of the flip-flops are at 1 when:

T_0 remains at 1 as it is connected V_{CC}.
$T_1 = 1$ when $U.Q_0 + D.Q_0' = 1$
$T_2 = 1$ when $U.Q_0.Q_1 + D.Q_0'.Q_1' = 1$
$T_3 = 1$ when $U.Q_0.Q_1.Q_2 + D.Q_0'.Q_1'.Q_2' = 1$

Counter output: $Q_3 Q_2 Q_1 Q_0$.

– Up-counter:
  The logic conditions could be simplified for up-counter. The input, U/D$'$ is set High for up-counting. Hence, U = 1 and D = 0. The logic conditions simplify to:
  $T_0 = 1$
  $T_1 = 1$ when $Q_0 = 1$
  $T_2 = 1$ when $Q_0.Q_1 = 1$
  $T_3 = 1$ when $Q_0.Q_1.Q_2 = 1$
  The above logic conditions are same as those obtained in Sect. 10.3.1.
– Down-counter:
  The input, U/D$'$ is set Low for down-counting. U = 0 and D = 1. The logic conditions simplify to:
  $T_0 = 1$
  $T_1 = 1$ when $Q_0' = 1$
  $T_2 = 1$ when $Q_0'.Q_1' = 1$
  $T_3 = 1$ when $Q_0'.Q_1'.Q_2' = 1$
  The up-down operation of the counter could be verified using state transition analysis as explained in earlier sections.

10.3.3.4 Reversing at Any Count Sequence

Up or down counting could be reversed at any state of count sequence using the input, U/D′. Assume that the up-down count sequence shown in Fig. 10.8a is required. The timing diagram for the up-down count sequence is shown in Fig. 10.8b. Input control (U/D′) and the output states of flip-flops are shown in the timing diagram.

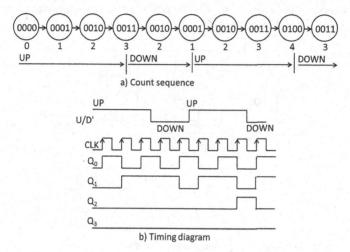

a) Count sequence

b) Timing diagram

Fig. 10.8 Reversing count sequence in 4-bit up-down Binary counter

10.4 Decoding Counter States and Glitches

The output states of counter need to be decoded in digital signal processing applications such as cascading counters. Decoding refers to obtaining active High or Low output signal for the desired counter output state. One or more AND or NAND gates are connected to the appropriate outputs of flip-flops for decoding. Decoding circuit is integral part of many counter ICs. For example, the Terminal count output of 74LS162 is High when the 4-bit Binary counter output is in its maximum count state i.e. 1111. Decoding gate circuits are presented for synchronous and asynchronous counters.

Glitches do not occur in the decoding circuits of synchronous counters as the flip-flops of the counters are triggered simultaneously. The datasheets of the ICs also mention that simultaneous clocking of flip-flops eliminates the output counting spikes that are normally associated with asynchronous counters. Glitches are associated with the decoding gates of asynchronous counters and they should be eliminated.

10.4.1 Decoder for Synchronous Counters

Synchronous 3-bit Binary counter with two decoding AND gates (A and B) is shown in Fig. 10.9. The AND gates output active High signals. The decoded output of AND gate, A is Z_7 and the AND gate goes High when the output state of the counter is 111. The decoded output of AND gate, B is Z_4 and the AND gate goes High Z_4 when the output state of the counter is 100. The AND gates could be replaced by NAND gates to obtain active Low signals.

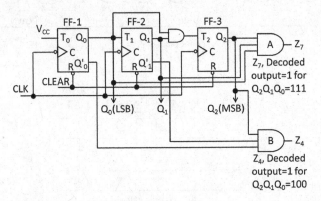

Fig. 10.9 Synchronous 3-bit Binary counter with decoder

10.4.1.1 Inputs for Decoding Gates

Decoding AND gates could be designed for one or all the eight states of the 3-bit Binary counter. Appropriate combination of the outputs of the flip-flops of the counter is used as the inputs for the AND gates. The combinations of the counter and decoder outputs are shown in Table 10.1 for the eight states of the 3-bit binary counter.

Table 10.1 Decoding counter outputs	Counter output	Inputs for decoding AND gate	Decoded output
	000	$Q_2'Q_1'Q_0'$	Z_0
	001	$Q_2'Q_1'Q_0$	Z_1
	010	$Q_2'Q_1Q_0'$	Z_2
	011	$Q_2'Q_1Q_0$	Z_3
	100	$Q_2Q_1'Q_0'$	Z_4
	101	$Q_2Q_1'Q_0$	Z_5
	110	$Q_2Q_1Q_0'$	Z_6
	111	$Q_2Q_1Q_0$	Z_7

10.4.1.2   Using Decoder ICs

Decoder ICs instead of basic gates are used for decoding counter output. For example, 4-to-10 Decoder IC, 74LS42, is used for decoding the outputs of BCD counter. The outputs of the counter are connected to the select inputs of the decode IC. The decoded outputs are available at the output ports of the decoder IC.

10.4.2   Decoder for Asynchronous Counters

The method of designing decoding AND gates for one or more states of asynchronous (ripple) counters is same as those for synchronous counters. However, glitches are associated with the outputs of the AND gates. One method of eliminating the glitches is presented.

10.4.2.1   Glitches in Decoded Outputs

Triggering of the flip-flops of asynchronous counter is sequential. The output of one flip-flop triggers the next flip-flop. The propagation delay in the flip-flops causes glitches at the output of decoding AND or NAND gate. Glitches are unwanted changes in the logic states of outputs for a short duration.

10.4.2.2   Eliminating Glitches

The outputs of counter are appropriately selected as the inputs of decoder ICs. The strobe input of ICs is used for eliminating glitches. The decoder outputs active High or Low signal after it is enabled by the strobe input.

The clock signal is used as strobe input. If the flip-flops of counter are positive edge-triggered, the negative level of clock signal is used as strobe input. If the flip-flops of counter are negative edge-triggered, the positive level of clock signal is used as strobe input. The strobe input provides adequate time for the glitches to disappear thus eliminating the glitches in decoded output.

10.5   Cascading Counters

A flip-flop is the basic 1-bit counter and it is Mod-2 counter. It has two binary count states, 0 and 1. The Mod-2 counter with one flip-flop functions as ($\div 2$) counter. Flip-flops are cascaded to obtain higher modulus counters. For example, four flip-flops are cascaded to obtain Mod-16 counter. The Mod-16 counter with four flip-flops functions as ($\div 16$) counter.

Digital signal processing applications require higher order counters such as Mod-100 and Mod-256 counters. Higher order modulus counters are obtained by cascading counters. Mod-100 counter is obtained by cascading two Mod-10 counters. Mod-160 counter is obtained by cascading Mod-16 and Mod-10 counters. Mod-256 counter is obtained by cascading two Mod-16 counters.

Cascading of counters could be designed using asynchronous or synchronous counters. Cascaded synchronous counters are characterized by glitch-free performance and higher operating speed compared to cascaded asynchronous counters. Standard ICs for synchronous counters have provision for cascading. Cascading synchronous counters is explained and illustrated with examples. Understanding terminal count of counters is fundamental for cascading counters.

10.5.1 Terminal Count for Cascading

Terminal count is the last state of counter. For example, the terminal count of 3-bit counter is 111. The terminal count of 4-bit Binary counter is 1111. The terminal count of 4-bit Decade counter is 1001. Terminal count is used for cascading counters to obtain higher order counters. Decoder circuit is used to obtain High output when the terminal count of counters is reached. The decoded output of counter is applied to the input for the next counter that needs to be cascaded. Three examples are provided for understanding terminal count.

10.5.1.1 Mod-4 Counter

Consider cascading two flip-flops i.e. two Mod-2 counters for realizing Mod-4 counter. As the terminal count of first flip-flop (FF-1) is 1, the output of FF-1 is directly connected to the input of FF-2 as shown in Fig. 10.10. Decoder is not needed.

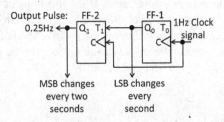

Fig. 10.10 Synchronous Mod-4 ($\div 4$) counter

Assume 1 Hz clock signal is used for edge-triggering the flip-flops. The input to FF-2 is available once in every two seconds from FF-1. Hence, the output state of FF-2 changes every two seconds. The output pulse of the second flip-flop appears once in every four seconds. The output trigger frequency is 0.25 Hz. The Mod-4 counter functions as ($\div 4$) counter.

10.5.1.2 Mod-16 Counter

Consider the synchronous 4-bit Mod-16 Binary counter, shown in Fig. 10.3. The Mod-16 counter is another example of cascading four flip-flops. Each flip-flop functions as Mod-2 counter. As the terminal count of first flip-flop (FF-1) is 1, the output of FF-1 is connected to the input of FF-2 directly.

The cascaded combination of FF-1 and FF-2 is Mod-4 counter. The terminal count of the Mod-4 counter is 11. Hence, the terminal count of the Mod-4 counter is decoded using AND gate, A, and applied to the input of FF-3.

The cascaded combination of FF-1, FF-2 and FF-3 is Mod-8 counter. The terminal count of the Mod-8 counter is 111. Hence, the terminal count of the Mod-8 counter is decoded using AND gate, B, and applied to the input of FF-4.

10.5.1.3 Mod-20 Counter

Mod-20 counter is obtained by cascading Mod-10 Decade counter and Mod-2 counter. Mod-2 counter has one flip-flop. Cascading Mod-10 Decade counter and Mod-2 counter is shown in Fig. 10.11a. The block diagram of Decade counter is shown with trigger input and outputs in the figure. The terminal count of the counter is 1001. It is decoded using AND gate and applied as input to the Mod-2 counter (flip-flop). The operation of the counter is explained.

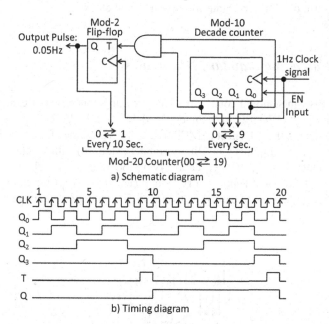

Fig. 10.11 Cascading Mod-10 and Mod-2 counters

10.5.1.4   Operation of Mod-20 Counter

Assume 1 Hz clock signal is used for triggering the Decade counter and the flip-flop. Initially, the output states of Decade counter and the flip-flop are reset to 0. The input of first flip-flop of the Decade counter should be set to 1 for the operation of the counter. The EN input, shown in Fig. 10.11a, sets the input of the first flip-flop of the counter to 1.

Clock signal triggers the Decade counter and the flip-flop every second. The output state of the Decade counter changes every second and the count sequence is 0000, 0001, 0010 and so on up to 1001. The output state of the flip-flop continues to remain at 0 until the Decade counter reaches its terminal count, 1001. The Decade counter reaches the terminal count, 1001, after 9 s. The terminal count is decoded by the AND gate and the output of the gate is 1. The input of the flip-flop becomes 1. After 10 s, the output of the flip-flop becomes 1 and the output of the Decade counter is reset to 0000. The decoded output from the Decade counter is 0 and hence the input to the flip-flop becomes 0. The output state of the flip-flop (Q) continues to remain at 1 until the Decade counter reaches again its terminal state, 1001. The timing diagram for the Mod-20 counter is shown in Fig. 10.11b.

If 7-Segment display is connected to the outputs of the flip-flop and the Decade counter, the combined decimal counting sequence, 00 to 19, of the Mod-20 counter would be displayed by the display. The display of Decade counter changes from 0 to 9 for every second. The display of flip-flop changes from 0 to 1 for every 10 s. Truncated form of Mod-20 counter is used in digital clock.

10.5.2   Cascading Standard IC Counters

Provision exists in standard counter ICs for cascading counters. For example, consider synchronous 4-bit Binary counter (Mod-16), MC74HC163A for cascading. The information in the datasheet of MC74HC163A is presented for cascading counters. The IC is a synchronous 4-bit Binary counter (Mod-16) with synchronous reset. Mod-256 counter is realized by cascading two Mod-16 counters. The IC has Count Enable T input (EN) and Ripple Carry Output (RC) provisions for cascading. The simplified block diagram of cascading two Mod-16 counters is shown in Fig. 10.12.

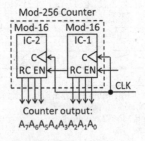

Fig. 10.12 Block diagram of cascaded Mod-256 counter

The input of the first flip-flop of IC-1 is set to 1 by Count Enable T input (EN). The Ripple Carry Output of first IC (IC-1) is connected to the Count Enable T input of second IC (IC-2). The Ripple Carry Output becomes 1 when IC-1 reaches its terminal count, 1111. The cascaded count sequence is from 00000000 to 11111111 with 256 states. The decimal equivalent of the count is from 0 to 255. The datasheet of the IC could be referred for the schematic diagram and other controls for counting.

10.5.2.1 Changing Counter Modulus

Counters that use the maximum number of output states are full modulus counters. For example, 4-bit Binary counters are full modulus counters (Mod-16) as they use all the 16 output states. The number of output states of counters could be limited or truncated to lower value using standard gates.

Realizing Decade counter from 4-bit Mod-16 Binary counter is an example of truncating the counter modulus to 10 with an add-on external logic gate circuit. Similarly, Mod-3 counter could be realized from 2-bit Mod-4 counter; Mod-5 to Mod-7 counters could be realized from 3-bit Mod-8 counter; Mod-9 to Mod-15 counters could be realized from 4-bit Mod-16 counter and so on. Digital applications require varying the modulus of counters.

Standard ICs have provisions for varying the modulus of counters. MC74HC163A is a programmable synchronous 4-bit Mod-16 Binary counter. External logic gates are added to the outputs of counters. Examples are available in the datasheet for varying the Mod-16 counter to Mod-5 and Mod-11 with minimal external logic gates. The output of the counters is glitch-free due to synchronous Reset.

10.5.2.2 Presetting Count Sequences

Presetting count sequences of counters is required for digital applications. For example, the count sequence of 4-bit Binary counter, MC74HC163A, is from 0000 to 1111. Assume that the count sequence, 1011-1100-1101-1110-1111-0000-0001-0010, is required at the output of the counter. Provision exists in the IC for presetting the count sequence.

The IC has Preset Data Inputs, $P_3P_2P_1P_0$, and Load control input. Initially, the counter outputs are reset to 0000. The data, 1011, is loaded into $P_3P_2P_1P_0$. A Low level is maintained for the load control input. With the next rising edge of the clock, the data from $P_3P_2P_1P_0$ appears at the output of the counter, $Q_3Q_2Q_1Q_0$. After loading the data, the load control is disabled and counting is enabled. The counter outputs the required count sequence. Counting is inhibited after the counter outputs 0010. The datasheet of the IC could be referred for the timing diagram of the count sequence. The term, Parallel Enable, instead of Load control input is also used in the datasheets of ICs.

10.6   Digital Clock

Digital clock is one of the applications of counters. The clock is commercially popular and it is installed in all public places. The clock displays time in hours, minutes and seconds. Generally, 7-segment displays are used for displaying time. Recently, Nixie tube display is being revived and it is capable of illuminating numbers, letters and symbols in scientific and industrial environments [1].

Counter is the basic device used in digital clock. The block diagram of 12-h clock using cascaded synchronous counters is shown in Fig. 10.13. Usually, digital clock has provision for users to set the current time. The provision is not shown in the figure. The basic operation of digital clock is presented.

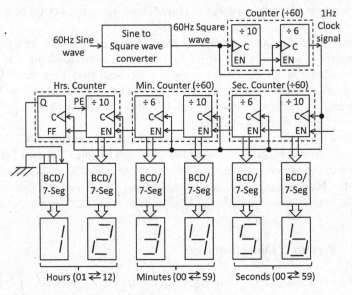

Fig. 10.13   Block diagram of 12-h digital clock

10.6.1   Operation

The basic analog input to digital clock is 60 Hz sine wave. The analog signal could be derived either from mains power or from crystal oscillator followed by a set of appropriate frequency divider circuit. The 60 Hz analog signal is converted into 60 Hz square wave (digital signal) using sine to square wave converter. The 60 Hz digital signal is down converted into 1 Hz signal by cascading ($\div 10$) counter and ($\div 6$) Decade counter. The signal path from 60 Hz analog signal to 1 Hz clock signal is shown in Fig. 10.13. The 1 Hz digital signal is the clock signal for triggering the set of cascaded counters for seconds, minutes and hours.

10.6.1.1 Cascaded Counters for Seconds

Digital clock displays seconds in two decimal digits from 00 to 59. The digits at ones position vary from 0 to 9 and hence Mod-10 Decade counter is required. The digits at tens position vary from 0 to 5 and hence Mod-6 counter is required. The two counters are cascaded. Enable input (EN Input) sets the input of first flip-flop of the Decade counter to 1. The operation of the cascaded counter is same as the cascaded counter explained in Sect. 10.5.1.3 except the flip-flop is replaced by Mod-6 counter. The output state of the Decade counter changes from 0 to 9 every second and that of the Mod-6 counter changes from 0 to 5 every 10 s. The equivalent decimal numbers of the output states of Mod-10 and Mod-6 counters are displayed in 7-Seg display. The Mod-60 seconds counter outputs one pulse every 60 s i.e. every minute and the pulse enables the next stage Mod-60 minutes counter.

10.6.1.2 Cascaded Counters for Minutes

The construction of minutes counter is similar to the seconds counter. Mod-60 minutes counter is constructed by cascading Mod-10 Decade counter and Mod-6 counter. The output state of the Decade counter changes from 0 to 9 every minute and that of the Mod-6 counter changes from 0 to 5 every 10 min. The equivalent decimal numbers of the output states of Mod-10 and Mod-6 counters are displayed in 7-Seg display. The Mod-60 minutes counter outputs one pulse every 60 min i.e. every hour and the pulse enables the next stage hours counter.

10.6.1.3 Cascaded Counters for Hours

Hours counter is the truncated form of Mod-20 counter. It is constructed by cascading Mod-10 Decade counter and a flip-flop. Additional gate circuit is used for truncation as all the twenty states of the cascaded counter are not needed for displaying hours in digital clock. State transition diagram is appropriate for identifying the states needed for hours counter. The states that are needed for displaying hours are shown in Fig. 10.14. Each state indicates the output of flip-flop and Decade counter for the displayed hour. The inputs for designing hours counter are obtained from the state transition diagram and the inputs are:

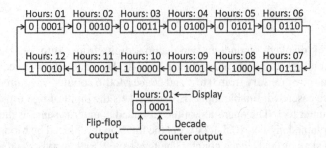

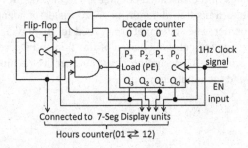

Fig. 10.14 Transition diagram of hours counter

(i)   The count sequence of the Decade counter should be preset to 0001.
(ii) The first count sequence of the Decade counter is from 0001 to 1001.
(iii) The next count sequence of the Decade counter should be truncated at 0010 and return to 0001.

10.6.1.4 Schematic Diagram of Cascaded Hours Counter

The schematic diagram of cascaded counter in Sect. 10.5.1.3 is modified for the hours counter and the modified diagram is shown in Fig. 10.15. The AND gate decodes the terminal count of the first count sequence of the Decade counter and provides the input for the flip-flop. When the second count sequence of the Decade counter starts, the output state of the flip-flop changes to 1 and it remains until the completion of the second sequence. When the count sequence of the Decade counter is 0010, the NAND gate decodes the output of the flip-flop and Q2 of the Decade counter. The output of the NAND gate provides active Low input to the Load (PE) of the Decade counter and loads the preset data, 0001 to the counter. Loading the preset data truncates the second count sequence of the Decade counter at 0010. The Sect. 10.5.2.2 could be referred for the details of loading preset data in counters.

Fig. 10.15 Cascaded hours counter

The output state of the Decade counter changes every hour and that of the flip-flop changes every 10 h. The equivalent decimal numbers of the output states of the Decade counter and the flip-flop are displayed in 7-Seg display.

10.7 Design of Counters

Counter design is converting the functional requirement of counter into simplified logic function. The logic function is implemented using appropriate digital hardware. The functional requirement is the count sequence of counter presented in the form of state transition diagram. The design of synchronous Mod-6 counter is presented. The counter is used in digital clock.

The count sequence of the Mod-6 counter used in digital clock has regular count sequence. Count sequence is considered regular if the decimal equivalent of the count sequence is incremented by one; otherwise, the count sequence is considered as irregular. Gray count sequence is an example of irregular count sequence. Some applications require irregular count sequence. The design of synchronous Mod-6 Gray code counter is also presented.

The general approach in Sect. 3.1.1 for designing combinational logic circuits is applicable for designing counters. The output states of the flip-flops of the counters are equivalent to the variables of combinational circuits. The truth table is replaced by state table. K-map is used to obtain the simplified logic functions for counter design. The logic functions relate the inputs and the outputs of the flip-flops. The procedure is illustrated for the design of the Mod-6 binary counter and Mod-6 Gray code counter using T flip-flops.

10.7.1 Mod-6 Binary Counter

The regular count sequences with equivalent decimal numbers of Mod-6 counter are shown in Fig. 10.16. The counter requires three flip-flops and they are labelled as FF-1, FF-2 and FF-3. The flip-flops are simultaneously triggered using common clock signal. The output state of FF-3 is the MSB and that of FF-1 is the LSB.

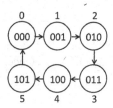

Fig. 10.16 Count sequences of Mod-6 binary counter

The inputs of the flip-flops (FF-1, FF-2 and FF-3) are T_0, T_1 and T_2. The present output states of the flip-flops are denoted as Q_0, Q_1 and Q_2 respectively. The next output states of the flip-flops are denoted as Q_0^+, Q_1^+ and Q_2^+. The state table of the counter is prepared relating the outputs and inputs of the flip-flops.

10.7.1.1 State Table

The state table has three major columns indicating the present output state, next output state and inputs of FF-1, FF-2 and FF-3. The present and next output states of the flip-flops are obtained from the count sequence of the Mod-6 counter, shown in Fig. 10.16. For example, consider the first present output state, 000. The next output state is 001. Similarly, the next output state for the present output state, 001, is 010. The present and next output states are shown in Fig. 10.17.

Count sequence	Present output state			Next output state			Inputs		
	Q_2	Q_1	Q_0	Q_2^+	Q_1^+	Q_0^+	T_2	T_1	T_0
0	0	0	0	0	0	1	0	0	1
1	0	0	1	0	1	0	0	1	1
2	0	1	0	0	1	1	0	0	1
3	0	1	1	1	0	0	1	1	1
4	1	0	0	1	0	1	0	0	1
5	1	0	1	0	0	0	1	0	1

Fig. 10.17 Output states and inputs of Mod-6 binary Counter

10.7.1.2 Deciding Flip-Flop Inputs

T flip-flop toggles output state if the input state is 1 before the arrival of clock trigger. If the input state is 0, the output state continues to remain in the same state. The toggling characteristic is used for deciding the inputs of the flip-flops. The inputs of the flip-flops are filled up individually considering the relevant present and next output states of the flip-flops. T_2 is filled up considering Q_2 and Q_2^+; T_1 is filled up considering Q_1 and Q_1^+; T_0 is filled up considering Q_0 and Q_0^+. The filling up of T_2, T_1 and T_0 is illustrated for the count sequences, 0, 2 and 4.

Count sequence-0:

$$Q_2 = 0 \text{ and } Q_2^+ = 0. \text{ Hence, } T_2 = 0$$
$$Q_1 = 0 \text{ and } Q_1^+ = 0. \text{ Hence, } T_1 = 0$$
$$Q_0 = 0 \text{ and } Q_0^+ = 1. \text{ Hence, } T_0 = 1$$

Count sequence-3:

$$Q_2 = 0 \text{ and } Q_2^+ = 1. \text{ Hence, } T_2 = 1$$
$$Q_1 = 1 \text{ and } Q_1^+ = 0. \text{ Hence, } T_1 = 1$$
$$Q_0 = 1 \text{ and } Q_0^+ = 0. \text{ Hence, } T_0 = 1$$

Count sequence-5:

$$Q_2 = 1 \text{ and } Q_2^+ = 0. \text{ Hence, } T_2 = 1$$
$$Q_1 = 0 \text{ and } Q_1^+ = 0. \text{ Hence, } T_1 = 0$$
$$Q_0 = 1 \text{ and } Q_0^+ = 0. \text{ Hence, } T_0 = 1$$

Similarly, the input states for the count sequences, 1, 2 and 4, could be determined. The input states for all the count sequences are shown in Fig. 10.17.

10.7.1.3 Logic Functions

The simplified logic functions for the combinational logic circuit relating the output and input states of flip-flops are obtained using K-maps. The data in Fig. 10.17 is split into three tables indicating the present output states and one input state for preparing K-map. T_2, T_1 and T_0 are the three tables and they are shown in Fig. 10.18a–c respectively. The K-maps for the tables are also shown in the figure. Don't care conditions (x) are shown in K-maps for unavailable output states. The simplified logic functions for T_2, T_1 and T_0 are also shown in the figure. K-map for T_0 is not necessary as $T_0 = 1$ for all count sequences.

Q_2	Q_1	Q_0	T_2
0	0	0	0
0	0	1	0
0	1	0	0
0	1	1	1
1	0	0	0
1	0	1	1

Q_2	Q_1	Q_0	T_1
0	0	0	0
0	0	1	1
0	1	0	0
0	1	1	1
1	0	0	0
1	0	1	0

Q_2	Q_1	Q_0	T_0
0	0	0	1
0	0	1	1
0	1	0	1
0	1	1	1
1	0	0	1
1	0	1	1

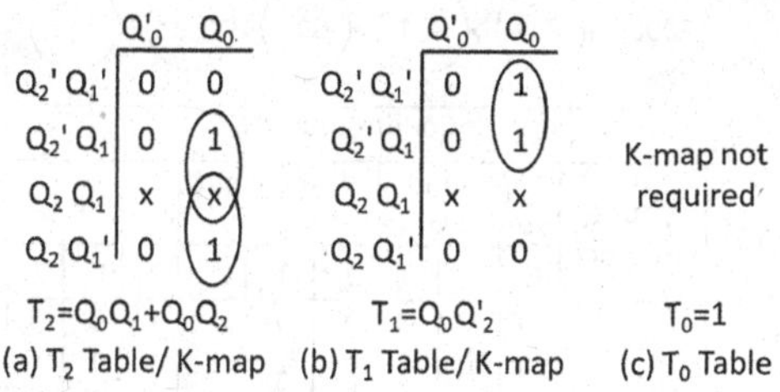

	Q_0'	Q_0
$Q_2'Q_1'$	0	0
$Q_2'Q_1$	0	1
Q_2Q_1	x	x
Q_2Q_1'	0	1

$T_2 = Q_0Q_1 + Q_0Q_2$

(a) T_2 Table/ K-map

	Q_0'	Q_0
$Q_2'Q_1'$	0	1
$Q_2'Q_1$	0	1
Q_2Q_1	x	x
Q_2Q_1'	0	0

$T_1 = Q_0Q_2'$

(b) T_1 Table/ K-map

K-map not required

$T_0 = 1$

(c) T_0 Table

Fig. 10.18 K-maps for Mod-6 binary counter

10.7.1.4 Implementation

The logic functions for T_2, T_1 and T_0 are shown in Fig. 10.18 for implementing Mod-6 counter. The implementation of synchronous Mod-6 counter is shown in Fig. 10.19. T_0 is connected to V_{CC}. The input gate circuits for T_2 and T_1 are as per the logic functions.

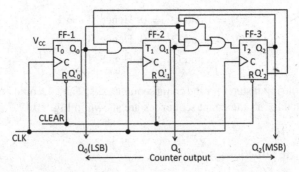

Fig. 10.19 Synchronous Mod-6 binary counter

10.7.2 *Mod-6 Gray Code Counter*

The method of designing Mod-6 counter with regular count sequence is applicable for Mod-6 Gray code counter. The count sequence, output and input states of flip-flops and K-maps for the inputs of T flip-flops are shown in Fig. 10.20 for Mod-6 Gray code counter. The equivalent decimal values of the Gray code count sequence are 0, 1, 3, 2, 6 and 7. Don't care conditions (x) are shown in K-maps for unavailable output states. The simplified logic functions for T_2, T_1 and T_0 are also shown in the figure.

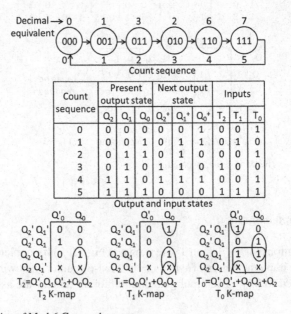

Fig. 10.20 Design of Mod-6 Gray code counter

10.7.2.1 Implementation

The logic functions for T_0 and T_1 are further simplified using XNOR function for implementation. The implementation of synchronous Mod-6 Gray code counter is shown in Fig. 10.21. The input gate circuits are as per the logic functions for T_2, T_1 and T_0.

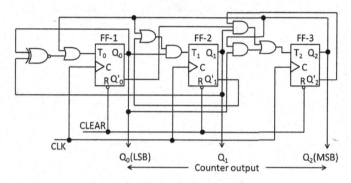

Fig. 10.21 Synchronous Mod-6 Gray code counter

$$T_0 = Q_0'Q_1' + Q_0Q_1 + Q_2 = (Q_0 \odot Q_1) + Q_2$$

$$T_1 = Q_0(Q_1' + Q_2)$$

Reference

1. Boos J (2018) The nixie tube story. IEEE Spectr 36–41

Chapter 11
Signal Converter Architectures

Abstract Analog signals are converted into digital signals for digital signal processing. Analog-to-Digital Converter (ADC) is used for the conversion process. When required, Digital-to-Analog Converter (DAC) is used for reconverting the processed digital signals into analog signals. Sample-Hold amplifier, Anti-aliasing filter, Reconstruction filter and Operational amplifier (Op-amp) are the analog components that are generally used in signal conversion circuits. The operation of the analog components is presented. Four ADC architectures, namely, Flash ADC, Dual slope ADC, Successive approximation ADC and Sigma-Delta ADC, are explained. Two DAC architectures, namely, Binary-weighted input DAC and R-2R Ladder DAC, are explained.

11.1 Overview of Signal Conversion

Electrical signals are processed for controls, monitoring, transmission and other requirements. The signals could be processed by analog or digital circuits. Digital signal processing has distinct advantages over analog signal processing. Higher accuracy in controls, reduced noise in transmission and encryption are some of the advantages. Section 1.4.1 could be referred for additional advantages.

Analog signals should be converted into digital signals for digital signal processing. Analog-to-Digital Converter (ADC) is used for the conversion process. When required, Digital-to-Analog Converter (DAC) is used for reconverting the processed digital signals into analog signals. If digital signals are used to drive digital displays, reconversion is not required; otherwise, reconversion is required. Pre-processing and post-processing requirements in signal conversion are presented.

11.1.1 Pre-processing and Post-processing

Analog signal could be DC voltage or continuously varying voltage (also current) with time. Analog signal is generally pre-processed using appropriate circuits before

© Springer Nature Switzerland AG 2020　　　　　　　　　　　　　　　　　225
D. Natarajan, *Fundamentals of Digital Electronics*,
Lecture Notes in Electrical Engineering 623,
https://doi.org/10.1007/978-3-030-36196-9_11

applying the signal to ADC for conversion. The output of the pre-processor is the input to ADC. The pre-processing circuit depends on the nature of analog signal.

For DC voltage measurements, the input signals are scaled by a combination of attenuators, DC Preamplifier and a low pass filter before presenting to ADC for conversion [1]. Generally, analog voltages that vary with time are the inputs for digital signal processing. They are pre-processed using analog components, namely, low pass filters, Sample-Hold Amplifier (SHA), Op-amp and other application specific circuits. Op-amp is also used in ADC and DAC circuits. The operation of the analog components is explained in Sect. 11.2.

After processing, digital signals are reconverted by DAC into analog signals. The output of DAC is passed through a filter for reconstructing the original analog signal. The simplified functional block diagram for digital signal processing (DSP) is shown in Fig. 11.1.

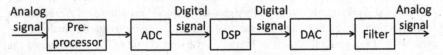

Fig. 11.1 Functional block diagram for digital signal processing

11.1.2 ADC

The overview of the operation of ADC is presented for converting DC voltage into digital signal for measurement. Pre-processing requirements of DC voltage is ignored for presenting the overview. Assume that DC voltage range is 0–7 V. It is also assumed that the voltage should be converted into 3-bit digital signal ($Q_2Q_1Q_0$). 3-bit ADC is required for the conversion and it has eight output states, 000 to 111. A reference input voltage to ADC is required for conversion and it is assumed as 8 V. The output binary code of the ADC is 000 for the input voltage range, ($V \leq 0.5$); it is 001 for the voltage range, ($0.5 < V \leq 1.5$); it is 010 for the voltage range, ($1.5 < V \leq 2.5$) and so on. The output code is 111 for the voltage range, ($6.5 < V \leq 7.5$). The functional block diagram of ADC and the output codes of the ADC for the voltage ranges are shown in Fig. 11.2a, b respectively. Quantization and quantization error are two fundamental terms associated with ADC. The terms are explained.

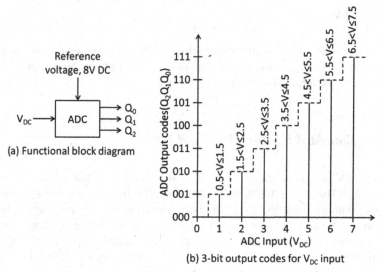

Fig. 11.2 Overview of ADC functioning

11.1.2.1 Quantization

Quantization is the process of converting discrete voltage levels of analog signal into binary codes. The basic function of ADC is quantization. The discrete voltage levels of analog signal are called quantization levels. Each quantization level is converted by ADC into digital code. There are eight quantization levels (0–7 V) for the DC voltage example in Sect. 11.1.2. They are converted into 3-bit code. Quantization interval between the levels is 1 V in the example and it is uniform.

11.1.2.2 Quantization Error

It could be observed in Fig. 11.2b that the output digital code of ADC is same for a range of input voltages. The range of voltages is indicated by horizontal dashed lines for each quantized level of input DC voltage in the figure. For example, the output code is 101 for the input voltage range, $(4.5 < V \leq 5.5)$. The output code is 110 for the input voltage range, $(5.5 < V \leq 6.5)$. Maximum error in the display is 0.5 V. The error is known as quantization error. It could be observed that one LSB represents 1 V and the voltage level transition occurs at 1/2 LSB. Quantization error is said to be 1/2 LSB.

One method of reducing quantization error is to increase the number of quantization levels. For the DC voltage example in Sect. 11.1.2, the quantization interval could be reduced to 0.5 V by increasing the number of quantized levels to sixteen. 4-bit ADC is required for the conversion. Maximum error in the digital voltmeter display reduces to 0.25 V. Additional information on ADC is available in Ref. [2].

11.1.3 DAC

DAC converts binary codes after digital signal processing into analog signal using appropriate architecture. Reconstruction filter is used at the output of DAC to smoothen the converted analog signal.

11.1.4 Standard ICs for ADCs and DACs

Standard ADC ICs are available in various configurations with many features. Examples are ADC0808, ADC0809, ADC1801 and AD9461. Minimizing or eliminating quantization errors and total harmonic distortion in conversion are some of the features. Some ICs have internal reference voltage. Some ICs have internal pre-processing circuit so that analog signal could be directly applied to the ICs.

Standard ICs are available for DACs also. Examples are DAC0800, DAC0802, DAC0808 and DAC5681. ICs are available with internal on-chip reference voltage. The datasheets of ADC and DAC ICs could be referred for additional details.

11.2 Analog Components for Signal Conversion

Analog components are used for pre-processing analog signals and for the operation of ADC and DAC. Four components, namely, Sample-Hold Amplifier (SHA), Anti-aliasing filter, Reconstruction filter and Operational amplifier (Op-amp) are the analog components that are generally used in signal conversion circuits. The construction and operation of the analog components are presented.

11.2.1 Sample-Hold Amplifier

The input signal to ADC should be in the form of discrete voltage levels for conversion into digital signal. Sample-Hold Amplifier (SHA) converts continuously varying analog voltage into discrete voltage levels. The conversion process is explained for sinusoidal signal and it is applicable for all types of analog signals that vary with time.

11.2.1.1 Functional Block Diagram

The functional block diagram of SHA is shown in Fig. 11.3. Clock-controlled switch and a capacitor at the output of the switch are the two basic components of SHA. In addition, the input and output of SHA are buffered by Op-amps to prevent loading effects.

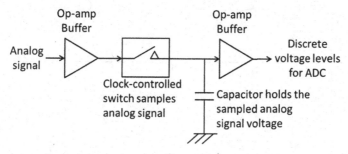

Fig. 11.3 Functional block diagram of SHA

The clock-controlled switch samples the input analog signal. The capacitor holds the sampled voltage of the analog signal. The input Op-amp is capable of delivering the required charging current to the hold capacitor. It also prevents loading effect to analog signal source, caused by the initial in-rush charging current of the capacitor. The output Op-amp prevents loading effect to the hold capacitor to retain the sampled voltage. ADC is connected to the output of SHA. It is not shown in the figure.

11.2.1.2 Operation

The operation of SHA is explained for one cycle of sinusoidal input signal, shown in Fig. 11.4. Assume that the signal shown in Fig. 11.4a is used for controlling the operation of clock-controlled switch. When the clock signal is Low, the switch is closed and the SHA is in sample (S) mode. The output capacitor of SHA tracks the input analog voltage in sample mode. When the clock signal is High, the switch is opened and the SHA is in hold (H) mode. The capacitor holds the analog voltage level at the instant of changing over from sample mode to hold mode. The operation of SHA is illustrated for two sampling points, t_0 and t_1.

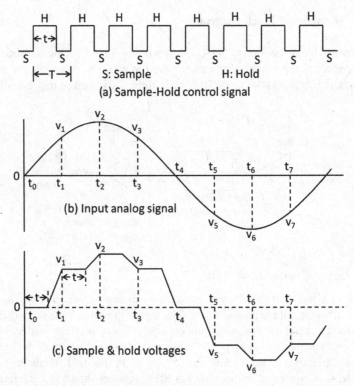

Fig. 11.4 Input and output signals of SHA

11.2.1.3 Illustrations

Consider the first sampling point, t_0, of the sinusoidal signal. SHA is in hold mode. The instantaneous voltage level of the analog signal at t_0 is 0 V. The output capacitor holds 0 V for the duration, t and it is shown by the horizontal line in Fig. 11.4c. During hold mode, the ADC connected to the output of SHA converts 0 V into appropriate digital signal. SHA changes to sample mode after the duration, t. The capacitor tracks the input analog voltage in sample mode until the next sampling point, t_1. The tracked voltage is shown by the slanted line in the figure.

SHA changes from sample mode to hold mode at the second sampling point, t_1, of the sinusoidal signal. The instantaneous voltage level of the analog signal at t_1 is v_1. The capacitor holds v_1 for the duration, t as shown by the horizontal line in the figure. During hold mode, ADC converts v_1 into appropriate digital signal. SHA changes to sample mode after the duration, t. The capacitor tracks the input analog voltage in sample mode until the next sampling point, t_2. The tracked voltage is shown by the slanted line in the figure. Sample and hold operations are repeated for the sampling points, t_2 to t_7.

11.2.1.4 Sampling Frequency

The frequency of clock signal in Fig. 11.4a is $(1/T)$ Hz where T is the period of the signal. It is called sampling frequency, f_s. It could be observed that T is equal to 1/8 of the period of the sinusoidal signal in Fig. 11.4b. Sampling frequency, $f_s = 8f_a$, where f_a is the frequency of the sinusoidal signal. Sampling frequency is expressed as the number of samples per second. If f_a is 1 Hz, f_s would be 8 Hz i.e. 8 samples per second. If f_a is 100 Hz, f_s would be 800 Hz i.e. 800 samples per second.

The number of discrete voltage levels (quantization levels) at the output of SHA increases with increase in sampling frequency. The accuracy of the conversion of analog signal into digital signal increases with increase in quantization levels. There is a minimum limit for sampling frequency to prevent the distortion of the original waveform of analog signal. The minimum limit for sampling frequency is derived from Nyquist criteria, which specifies that the sampling frequency should be at least twice the highest frequency contained in the analog signal. In practice, sampling frequency is usually much higher for all applications.

11.2.1.5 Errors in Sampling Analog Signal

In addition to quantization error, time related and other types of errors occur in sample mode and hold mode. Errors also occur during transition from sample to hold and from hold to sample. Examples of the errors are droop rate, settling time, offset, aperture jitter and acquisition time. The errors are not shown in the input and output signal waveforms in Fig. 11.4. Limits for various types of errors are specified in the datasheets of the standard ICs for SHAs. Application notes of device manufacturers could be referred for additional information on the types of errors [3–5].

11.2.2 Anti-aliasing Filter

Non-sinusoidal analog waveforms are used in many digital signal processing applications. Analog signal could be audio or video or other types of signals, which are not sinusoidal. Consider voice signal band for telephone applications. The frequency band of voice signal ranges from 300 to 3400 Hz approximately. Assume that the sampling frequency is selected considering the maximum frequency of 3400 Hz and complying with Nyquist criteria for digital signal processing. However, there could be distortion in the reconstructed voice signal after digital signal processing. The voice signal might become indistinguishable. The errors are caused by the harmonic signal content in the input voice signal and frequencies higher than 3400 Hz. They are the replica of original analog signal at higher frequencies. The errors are called aliasing errors. Hence, the input analog signal is passed through anti-aliasing filter before applying the signal to SHA.

Anti-aliasing filter is analog low pass filter. The pass band of the low pass filter is the frequency band of analog signal. Pass band signals are allowed with minimum attenuation and the frequencies higher than the upper limit of the pass band are attenuated. For example, the pass band of the low pass filter for the voice band for telephone applications is (300–3400) Hz. Rejection requirements are specified for the frequencies above 3400 Hz.

11.2.3 Reconstruction Filter

The output of DAC is analog signal. Random variations might exist in the analog signal and they are caused by noise signals. The random variations in the analog signal are smoothened by using a low pass reconstruction filter. ITU (International Telecommunication Union) standards are available specifying the requirements of reconstruction filter for various applications. Texas application report [6] could be referred for the design and operation of analog reconstruction filter video applications.

11.2.4 Operational Amplifier

The invention of Operational amplifier (Op-amp) dates back to early 20th century in vacuum tube era. The IC version of Op-amp was invented in 1958 by Jack Kilby of Texas Instruments [7]. The symbol of Op-amp with input and output connections is shown in Fig. 11.5. Op-amp has two signal inputs, namely, inverting signal input ($-V_{in}$) and non-inverting signal input ($+V_{in}$). The signal output of Op-amp is available at V_{out}. Op-amp requires dual power supply, namely, positive supply voltage ($+V_{supply}$) and negative supply voltage ($-V_{Supply}$). Op-amp has one more input, namely, Offset null, and the input is used to eliminate the offset voltage and balance the input voltages. Offset null input is not shown in the figure. Op-amp is used in three functional configurations in signal conversion circuits.

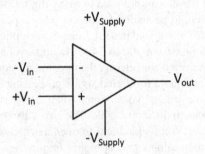

Fig. 11.5 Op-amp with input and output connections

11.2.4.1 Op-Amp as Amplifier

Op-amp could be configured as non-inverting and inverting amplifiers. Signal is applied to the non-inverting input of Op-amp in non-inverting amplifier. Signal is applied to the inverting input in inverting amplifier. Both the configurations of Op-amps are used in signal conversion circuits.

Non-inverting Op-amps are used as buffer amplifier with high gain or as buffer with unity gain. Op-amp with unity gain is called voltage follower or simply Op-amp buffer. The configurations of Op-amp buffer amplifier and Op-amp buffer are shown in Fig. 11.6a, b respectively. Both the configurations of non-inverting Op-amps have high input impedance and low output impedance. The type of Op-amp is decided considering the requirement of signal conversion circuits. Op-amp buffer is used at the input and output of SHA. It is also used at the output of DAC.

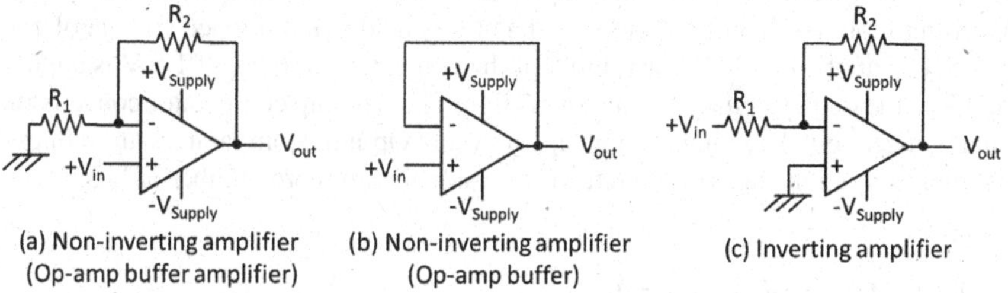

(a) Non-inverting amplifier
(Op-amp buffer amplifier)

(b) Non-inverting amplifier
(Op-amp buffer)

(c) Inverting amplifier

Fig. 11.6 Configurations of Op-amp as amplifier

Inverting Op-amps are used in DACs. The input impedance of the Op-amps is equal to the series resistance with input signal and the output impedance is low. The configuration of inverting amplifier is shown in Fig. 11.6 (c).

11.2.4.2 Op-Amp as Comparator

High speed Op-amps (Ex.: AD8057) are used as comparator in ADCs for signal conversion. The general representation of Op-amp as comparator is shown in Fig. 11.7a. If the non-inverting input ($+V_{in}$) is greater than the inverting input ($-V_{in}$), the output of Op-amp comparator is High; otherwise, the output is Low. The conditional outputs of Op-amp comparator are shown in the figure.

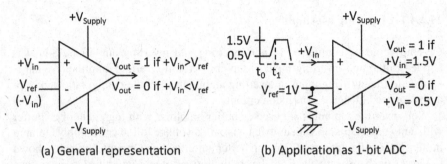

(a) General representation   (b) Application as 1-bit ADC

Fig. 11.7 Op-amp as comparator

The Op-amp comparator could be considered as 1-bit ADC. The schematic diagram of Op-amp comparator as 1-bit ADC is shown in Fig. 11.7b. The discrete voltage level waveform input to the ADC is shown in the figure. For example, assume that the inverting input $(-V_{in})$ of Op-amp comparator is held at a reference voltage of 1 V. If 0.5 V is applied to $+V_{in}$, the output of the comparator is Low. If 1.5 V is applied to $+V_{in}$, the output of the comparator is High. The Op-amp comparator converts the analog discrete voltage into digital signal. Many Op-amp comparators are required for representing the analog discrete voltage signal with more number of bits.

11.2.4.3   Op-Amp as Integrator

The schematic diagram of Op-amp as integrator is shown in Fig. 11.8a. The relationship between the input and output of Op-amp is given by,

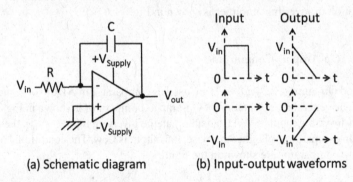

(a) Schematic diagram   (b) Input-output waveforms

Fig. 11.8 Op-amp as integrator

$$V_{out} = -\frac{1}{RC} \int_0^t V_{in}(t)dt$$

The input and output waveforms of the Op-amp integrator for $+V_{in}$ and $-V_{in}$ are shown in Fig. 11.8b. If V_{in} is positive voltage, the output is a ramp with negative slope falling from $+V_{in}$ to 0 V. If V_{in} is negative voltage, the output is a ramp with positive slope rising from $-V_{in}$ to 0 V. The magnitude of the slope of the ramp is (V_{in}/RC).

11.3 ADC Architectures

Many types of architectures have evolved over a period of time for converting pre-processed analog signals into binary codes considering speed and performance requirements. Four types of architectures are presented using simplified functional block diagrams. The types of architectures are:

(i) Flash ADC
(ii) Dual slope ADC
(iii) Successive approximation ADC
(iv) Sigma-Delta ADC.

11.3.1 Flash ADC

Flash ADC converts analog discrete voltages into digital signals quite fast. The ADC requires $(2^n - 1)$ comparators with input analog signal having 2^n discrete voltage levels i.e. quantization levels. 3-bit flash ADC requires $(2^3 - 1)$ i.e. seven comparators. It is capable of converting 2^3 i.e. eight quantization levels of analog signal into digital signals.

11.3.1.1 Functional Block Diagram

The simplified schematic diagram of 3-bit flash ADC is shown in Fig. 11.9. Op-amp comparators (1–7) and priority encoder with latch are shown in the figure. The waveforms of positively varying input analog discrete voltage signal to the ADC and priority encoder enable signal (E_{in}) are also shown in the figure. The discrete voltage level varies from 0.5 to 7.5 V. 3-bit unipolar straight binary code is suitable to represent the converted digital signal.

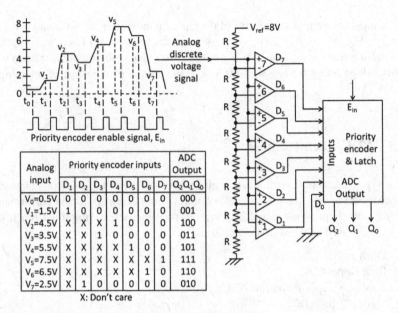

Fig. 11.9 Simplified schematic diagram of 3-bit flash ADC

Reference voltage is required for Op-amp comparators for converting the input discrete voltage levels into 3-bit codes. A resistor divider network with 8 V DC is used to apply reference voltage to each comparator. The reference voltage to the Comparator-1 is 1 V; it is 2 V for the Comparator-2 and so on. The output of flash ADC is $Q_2Q_1Q_0$ and it is available at the output of the priority encoder. The output digital signal is latched by the priority encoder enable signal for reading the signal.

The priority encoder enable signal (E_{in}) should be in synchronous with the clock signal of SHA. Synchronizing the signal ensures that all the discrete voltage levels of the input analog signal are converted into digital signals; otherwise, some of the voltage levels might be lost during conversion. Tutorial of Analog Devices could be referred for additional information on flash ADC architecture [8].

11.3.1.2 Operation

The operation of flash ADC is explained for converting three input discrete voltage levels into 3-bit codes. Consider the voltage level of the analog discrete voltage waveform at t_0 in Fig. 11.9. The discrete voltage level at t_0 is 0.5 V (V_0). When 0.5 V is applied as input to the flash ADC, the outputs of all comparators remain Low. The latched output of the ADC is 000.

The discrete voltage level at t_1 is 1.5 V (V_1). When 1.5 V is applied as input to the flash ADC, the output (D_1) of the comparator-1 goes High. The latched output of the ADC is 001.

The discrete voltage level at t_2 is 4.5 V (V_2). When 4.5 V is applied as input to the flash ADC, the outputs, D_1, D_2, D_3 and D_4 of the comparators go High.

The inputs, D_1, D_2 and D_3 are don't care inputs to priority encoder. The output of the priority encoder corresponds to the input, D_4, and the output of the ADC is 100. The conversion process continues for the discrete voltage levels at t_3, t_4, t_5, t_6 and t_7. The outputs of the ADC are tabulated in Fig. 11.9.

Analog discrete voltage signal varying with positive and negative voltages could also be converted into digital signal by applying appropriate reference voltages to the Op-amp comparators of flash ADC. Bipolar codes are used to represent the digital signals.

11.3.2 Successive Approximation ADC

Successive approximation ADC is characterized by high speed. Typically, the ADC is capable of performing 100,000 conversions per second [9]. The architecture is explained using 3-bit successive approximation ADC. Standard ICs are available for the architecture with 8-bits and higher.

The simplified schematic diagram of 3-bit successive approximation ADC is shown in Fig. 11.10. The ADC has Op-amp comparator, 3-bit Successive Approximation Register (SAR), Digital-to-Analog Converter (DAC) and Clock and Control unit. The closed loop interconnection between SAR, DAC and the Op-amp comparator is shown in the figure. The closed loop circuit performs successive comparisons with input analog signal for converting the analog signal into digital signal. The SAR contains three positive edge-triggered D flip-flops. The start and stop for the operations are controlled by the Clock and Control unit.

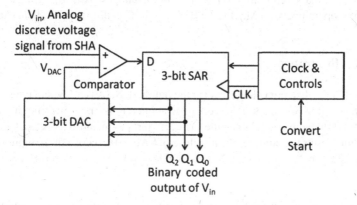

Fig. 11.10 Simplified diagram of 3-bit Successive approximation ADC

11.3.2.1 Operation

V_{in} is the analog discrete voltage from SHA and it is applied to the non-inverting input of Op-amp comparator. The output of the comparator is applied as the data input, D, to 3-bit Successive Approximation Register (SAR).

When the Convert Start command is asserted, Clock and Control unit resets the inputs of the flip-flops of SAR to 0 except that the input of MSB flip-flop is set to 1. The clock signal is also enabled. When the next positive transition (say trigger-1) of the clock signal appears, the flip-flops are triggered. The output of SAR is $Q_2Q_1Q_0$ and it is converted into analog output by the DAC. The analog output of the DAC is V_{DAC} and it is compared with V_{in}. If V_{in} is less than V_{DAC}, the output of the comparator is Low. The input of MSB flip-flop of SAR is reset to 0. If V_{in} is greater than V_{DAC}, the output of the comparator is High. The input of MSB flip-flop is kept as 1.

The Control unit sets the input of middle flip-flop to 1. After the next positive transition (trigger-2) of the clock signal, the analog output of the DAC (V_{DAC}) is compared with V_{in}. If V_{in} is less than V_{DAC}, the output of the comparator is Low. The input of middle flip-flop of SAR is reset to 0. If V_{in} is greater than V_{DAC}, the output of the comparator is High. The input of the middle flip-flop is kept as 1.

Finally, the input of LSB flip-flop of SAR is set to 1. After the next positive transition (trigger-3) of the clock signal, the analog output of the DAC (V_{DAC}) is compared with V_{in}. The input of the LSB flip-flop is reset to 0 or kept as 1 based on the results of comparison.

Clock trigger-4 is applied to obtain the final 3-bit output code. The clock signal is inhibited after trigger-4. The final output of the SAR in parallel format is $Q_2Q_1Q_0$ and it is the binary representation of the input analog signal, V_{in}. The successive comparing operation from MSB to LSB of the ADC is illustrated with an example.

11.3.2.2 Illustration

Assume that the discrete voltage level at the output of SHA is 3.5 V and it needs to be represented using 3-bit code. The voltage is applied to the non-inverting input (V_{in}) of the Op-amp comparator. When the Convert Start command is asserted, Clock and Control unit resets the inputs of the flip-flops of SAR to 0 except that the input of MSB flip-flop is set to 1. The clock signal is enabled. When the next positive transition (trigger-1) of the clock signal appears, the output of SAR ($Q_2Q_1Q_0$) becomes 100. The output of SAR is converted into analog output by the DAC. The equivalent analog output of the DAC (V_{DAC}) is 4 V. As V_{in} is less than V_{DAC}, the output of comparator is Low. The input of MSB flip-flop is set to 0.

The input of the middle flip-flop of SAR is set to 1. After the next positive transition (trigger-2) of the clock signal, the input to DAC is 010. The equivalent analog output of the DAC (V_{DAC}) is 2 V. As V_{in} is greater than V_{DAC}, the output of comparator becomes High. The input of the middle flip-flop is kept as 1.

The input of the LSB flip-flop of SAR is set to 1. After the next positive transition (trigger-3) of the clock signal, the input to DAC is 011. The equivalent analog output of the DAC (V_{DAC}) is 3 V. As V_{in} is greater than V_{DAC}, the output of comparator becomes High. The input of the LSB flip-flop is kept as 1.

Clock trigger-4 is applied to obtain the final 3-bit output code. The clock signal is inhibited after trigger-4. The final output code of SAR is 011. It is the binary representation of the input analog voltage, 3.5 V, accepting quantization error. The successive comparing operations are shown in Fig. 11.11.

V_{in}=3.5V

Clock Trigger	SAR FF Inputs before edge-triggering ($D_2D_1D_0$)	SAR Output after edge-triggering ($Q_2Q_1Q_0$)	V_{DAC}	Comparator output	Decision
Trig-1	MSB FF D_2 set to 1 100	100	4V	Low	Reset MSB to 0
Trig-2	MSB FF D_2 set to 0 & Middle FF D_1 set to 1 010	010	2V	High	Keep middle bit
Trig-3	Middle FF D_1 kept as 1 & LSB FF D_0 to 1 011	011	3V	High	Keep LSB
Trig-4	LSB FF D_0 kept as 1 011	011			

Binary representation of 3.5V: 011

Fig. 11.11 Conversion operations in 3-bit successive approximation ADC

11.3.3 Dual Slope ADC

DC voltage measurement with high accuracy and resolution is one of the important applications of dual slope ADCs. DC voltage could be from sensor interfaces such as thermocouples or generated voltages from current and resistance measurements.

The functional block diagram of Dual slope ADC is shown in Fig. 11.12. Dual slope ADC has analog and digital sections. The analog section of the ADC consists of switch, Op-amp integrator and Op-amp comparator. The digital section of the ADC consists of control logic, counter and digital display unit. Control logic unit obtains signals from comparator, clock and counter. The unit controls the operation of the switch for selecting input signals and counter. Counter output drives digital display unit. The converted analog input signal is displayed by the digital display unit.

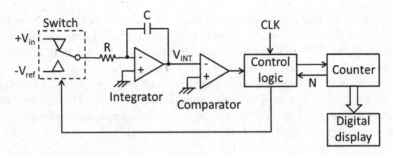

Fig. 11.12 Simplified schematic diagram of dual slope ADC

11.3.3.1 Operation of Dual Slope ADC

Integrating the input analog voltage of dual slope ADC is the key function in signal conversion. Standard ICs (Ex. ICL7135C and TC500) are available for dual slope ADC. Adequate information is available in the datasheets of device manufacturers explaining the design and operation of Op-amp integrator for signal conversion. V_{in} is the input DC voltage that needs to be converted. It could be positive or negative voltage. Another reference input voltage, V_{ref}, is also required for the operation of dual slope ADC. If V_{in} is positive, then V_{ref} should be negative and vice versa. The operation of dual slope ADC is explained assuming V_{in} is positive. IC datasheets specify the maximum and recommended full scale limits for V_{in} (positive and negative). The input DC voltage (V_{in}) should be within the recommended limits.

Op-amp integrator transforms input DC voltage ($+V_{in}$) into time measurement, proportional to the magnitude of the input voltage. The transformation is done in two stages, namely, charging and discharging. $+V_{in}$ is used for charging stage. The negative reference voltage ($-V_{ref}$) is the input voltage for discharging stage. Control logic unit provides signals for changing over from charging stage to discharging stage. The transformed time measurement is converted into digital signal by the counter and associated logic controls during discharging stage.

11.3.3.2 Charging Stage

The capacitor, C, of Op-amp integrator is charged during charging stage. Initially, the capacitor, C, of the Op-amp integrator is in discharged state and the comparator output is Low. The counter output is reset to 0. The duration for the charging stage is called the charge time, T_{chrg}. It is a fixed time period for all values of input DC voltages (V_{in}). Charge time is selected in the multiples of the period of line frequency (50 or 60 Hz) to maximize rejection of line frequency interference signals. If T_{chrg} is decided as two times the period of the line frequency, 50 Hz, the charge time would be 40 ms.

Assume that control logic unit provides signal to switch unit and $+V_{in}$ is selected. Charging stage begins after selecting $+V_{in}$. The capacitor (C) of the Op-amp integrator starts charging. When the capacitor starts charging, negative voltage is applied to the inverting input of Op-amp comparator. The non-inverting input of the comparator is at 0 V since it is grounded. As the non-inverting input voltage is greater than the inverting input voltage, the comparator output becomes High during the charging stage. The logic High signal enables control logic unit. The control logic unit enables clock signal to counter. The counter starts counting the pulses of the clock signal.

The counting operation of the counter should be stopped when charge time, T_{chrg}, elapses. The control logic unit monitors the counting of clock pulses. When the counter counts predetermined N number of clock pulses in the charge time (T_{chrg}), the control unit disables clock signal to counter. Counting operation is terminated. The predetermined N number of clock pulses is obtained from the expression,

$$N = f * T_{chrg}$$

T_{chrg} is in seconds and f is the clock frequency in Hz. If T_{chrg} is 40 ms and the clock frequency is 1 MHz, N is 40,000. The counting operation would be stopped by the control unit when the counter counts 40,000 clock pulses. Simultaneously, the control unit provides signal to Switch unit for selecting $-V_{ref}$ and the counter is reset to 0. The charging stage is completed.

The output voltage (V_{INT}) of the Op-amp integrator is a ramp from 0 V to $-V_{in}$ with negative slope, V_{in}/RC. The output waveform of charging stage is shown in Fig. 11.13. If the input voltage is reduced to 0.5 V_{in}, V_{INT} voltage would be a ramp from 0 V to -0.5 V_{in} with negative slope, 0.5 V_{in}/RC. T_{chrg} remains constant irrespective of the level of input voltage; only the slope of the ramp changes.

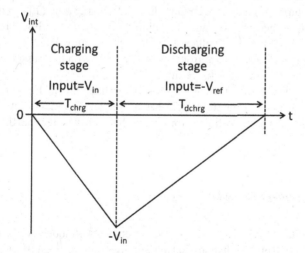

Fig. 11.13 Output waveforms of charging and discharging stages

11.3.3.3 Discharging Stage

Charge gained by the capacitor, C, Op-amp integrator during charging stage is discharged during discharging stage. The discharging stage begins after $-V_{ref}$ is selected by the Switch unit. When the input voltage ($-V_{ref}$) is applied to Op-amp integrator, the capacitor, C, of the Op-amp integrator starts discharging from $-V_{in}$. The comparator output remains High until the capacitor discharges to 0 V. When the comparator output is High, control logic unit enables clock signal to counter. The counter starts counting the pulses of the clock signal from the reset state. When the capacitor voltage becomes zero, the comparator output becomes Low. The Logic control unit disables clock signal. The counter output at the instant of disabling clock signal is latched.

The output voltage (V_{INT}) waveform of the Op-amp integrator is also a ramp from $-V_{in}$ to 0 V with positive slope, V_{ref}/RC. The output waveform of discharging stage is shown in Fig. 11.13. The duration of discharging is called the discharge time, T_{dchrg}. The discharging stage is completed.

The charge gained by the capacitor of Op-amp integrator during charging stage and the charge lost by the capacitor during discharging stage are equal. The mathematical equation relating charge gained and charge lost is,

$$\frac{V_{in}}{RC}T_{chrg} = \frac{V_{ref}}{RC}T_{dchrg}$$

$$T_{dchrg} = \frac{V_{in}}{V_{ref}}T_{chrg}$$

As T_{chrg} and V_{ref} are constants, T_{dchrg} is proportional to the input voltage, V_{in}. The latched output of the binary counter during discharge time (T_{dchrg}) is the binary representation of the input voltage. The digital display unit in Fig. 11.12 displays the input voltage. If the input voltage is $0.5V_{in}$, the output voltage (V_{INT}) of the Op-amp integrator would be a ramp from $-0.5V_{in}$ to 0 V with the same positive slope, V_{ref}/RC and having lower discharge time.

The output waveform of Dual slope ADC architecture in Fig. 11.13 has two slopes. Multi-slope ADC architectures are available. The multi-slope integrating ADC is capable of extremely high resolution with linearity better than 0.1 ppm of full scale [1].

11.3.4 Sigma-Delta ADC

The sigma-delta (Σ-Δ) ADC is the converter choice for modern voice band, audio and high resolution precision industrial measurement application [10]. The simplified schematic diagram of first order Σ-Δ ADC is shown in Fig. 11.14. The loop containing the Op-amp difference amplifier, Op-amp integrator, 1-bit Op-amp comparator and 1-bit DAC is the key functional block of the converter. The functional block is called Σ-Δ modulator. The analog signal that needs to be digitized is connected to the non-inverting input of the differential amplifier. The converted digital signal is available at the output of the comparator. The signal is passed through digital filters for obtaining final digital signal.

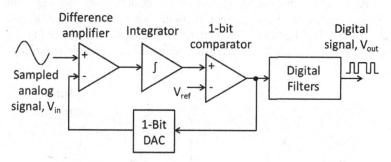

Fig. 11.14 First order sigma-delta ADC

The Σ-Δ ADC contains quite complex computational circuitry consisting of a digital signal processor [10]. Only the overview of the operation of the ADC is presented.

11.3.4.1 Operation

Sigma-Delta ADC uses the technique of oversampling for converting analog signal into high-resolution digital signal. Many samples of analog signal are made available as inputs to the ADC. The amplitude of the analog signal should vary slowly with time to facilitate oversampling.

The sampled input analog signal (V_{in}) and the analog output signal of 1-bit DAC are differentiated by the Op-amp difference amplifier. The output of the difference amplifier is applied to the Op-amp integrator. The output of the integrator is a ramp signal with positive or negative slope. The magnitude and the nature of the integrator output depend on the sign and magnitude of the output of the difference amplifier output. When the output voltage of the integrator is equal to V_{ref} of Op-amp comparator, the comparator output state changes. If the initial output state of comparator is 0, the output state changes to 1. If the initial state is 1, the state changes to 0. The output of the comparator is fed-back to the 1-bit DAC and looping cycle continues. The clock signal that is used for sampling analog signal controls the operation of Op-amp comparator also. The Σ-Δ modulator outputs 1-bit stream of digital signal and the ratio of the number of ones to zeros represents the input analog voltage [11].

11.3.4.2   Higher Order Sigma-Delta ADC

Higher order Σ-Δ ADCs are generally used for converting analog signal into digital signal. The higher order ADCs reduces quantization noise. Texas Instruments delta-sigma converters include second- through sixth-order modulators [11]. Higher order Σ-Δ ADCs is constructed with additional loops containing 1-bit DAC, Op-amp difference amplifier and integrator. The construction of second order Σ-Δ ADC is shown in Fig. 11.15a. The simplified representation of the ADC is shown in Fig. 11.15b. References [10–13] could be referred for additional information regarding Σ-Δ ADCs.

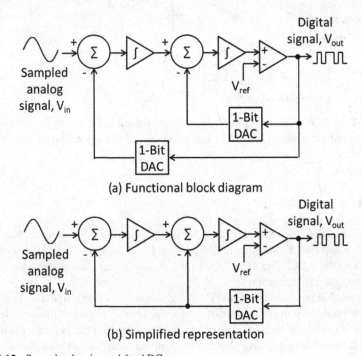

(a) Functional block diagram

(b) Simplified representation

Fig. 11.15   Second order sigma-delta ADC

11.4 DAC Architectures

DAC converts binary signal into discrete levels of analog voltages. The construction and operation of two types of DAC architectures are presented. The architectures are Binary-weighted input DAC and R-2R Ladder DAC.

11.4.1 Binary-Weighted Input DAC

Inverting Op-amp is the basic element in Binary-weighted DAC. It is used as summing amplifier. The output of the summing Op-amp is the weighted sum of the inputs. The construction and operation of Binary-weighted input DAC is presented after explaining the summing Op-amp.

11.4.1.1 Summing Op-Amp

Summing Op-amp is inverting Op-amp with multiple inputs. The schematic diagram of summing Op-amp is shown in Fig. 11.16. The current into the inverting input $(-)$ of the Op-amp is negligible. The inverting input is at 0 V and it is considered as virtual ground. Hence, the input currents through the three resistors are summed up and the resulting current flows through the feedback resistance, R_f. The output voltage of summing Op-amp (V_{out}) is the product of R_f and the weighted sum of the input currents. The direction of current through R_f depends on the polarity of input voltage. V_{out} is 180° out of phase with input voltage. The expression for V_{out} is indicated below. Binary-weighted input DAC is an application of the summing Op-amp.

$$V_{out} = -R_f \left(\frac{V_1}{R_1} + \frac{V_2}{R_2} + \frac{V_3}{R_3} \right)$$

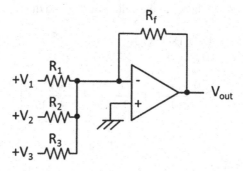

Fig. 11.16 Summing Op-amp

11.4.1.2 Operation

The schematic diagram of 3-bit Binary-weighted DAC is shown in Fig. 11.17. The input binary signal is represented by $D_2 D_1 D_0$. Three switches are shown with open status in the figure and it represents the binary signal, 000. If D_0 switch is closed, it represents 001. Closing of D_1 switch represents 010; closing of D_0 and D_1 switches represent 011 and so on. If all the three switches are closed, it represents 111.

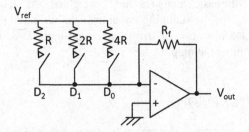

Fig. 11.17 Schematic diagram of 3-bit Binary-weighted input DAC

There are also resistors in series with the switches. The values of the resistors are selected in multiples of two as R, 2R and 4R. The resistor, R, is assigned to MSB (D_2) and it is in series with the MSB switch. 2R is assigned to the next bit (D_1) and it is in series with the next bit switch. 4R is assigned to LSB (D_0) and it is in series with the LSB switch. A reference input voltage (V_{ref}) is required for signal conversion and is shown in Fig. 11.17. V_{out} is the analog output voltage of the DAC.

$$V_{out} = -R_f \left[\frac{D_2 V_{ref}}{R} + \frac{D_1 V_{ref}}{2R} + \frac{D_0 V_{ref}}{4R} \right]$$

D_2, D_1 and D_0 are the status of input digital signal and they are either 0 or 1. V_{out} could be computed for all the eight states of 3-bit digital signal. Different values of V_{out} could be realized by changing the values of V_{ref} and the resistors.

Example:
Let $V_{ref} = -5$ V and $D_2 D_1 D_0 = 011$
$R_f = 8$ KΩ and R = 10 KΩ

$$V_{out} = -8 \left[\frac{0 * (-5)}{10} + \frac{1 * (-5)}{2 * 10} + \frac{1 * (-5)}{4 * 10} \right] = 3 \text{ V}$$

11.4.1.3 Limitation

The number of series resistors in binary-weighted input DAC is equal to the number of bits in the input digital signal. It could be observed that the values of resistors increase in the multiples of 2. For example, the LSB series resistor should be 8R for 4-bit signal; 16R for 5-bit signal and so on. The DAC architecture is not suitable for the conversion of higher order digital signals.

11.4.2 R-2R Ladder DAC

The input resistor network of R-2R DAC is modified to eliminate resistor values higher than 2R. Hence, the architecture is more popular and standard ICs (Ex. DAC8222) are also available. More number of resistors is required for R-2R Ladder DAC architecture compared to Binary-weighted input DAC. The schematic diagram of 3-bit R-2R Ladder DAC is shown in Fig. 11.18.

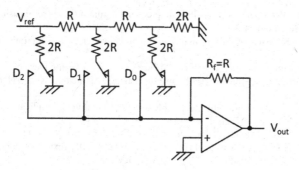

Fig. 11.18 Schematic diagram of 3-bit R-2R DAC

The input binary signal is represented by $D_2D_1D_0$. Three switches are shown with grounded status in the figure and it represents the binary signal, 000. They are CMOS transistor switches. Resistors are shown in series with the switches. Thin film resistors are used in the fabrication of ICs. A reference voltage, V_{ref}, is connected to the resistor network as shown in Fig. 11.18. If the reference voltage is switched to D_0 only, it represents 001. Switching the reference voltage to D_1 only represents 010; switching the reference voltage to D_0 and D_1 represent 011 and so on. If the reference voltage is switched to D_0, D_1 and D_2, it represents 111.

D_2, D_1 and D_0 are the status of input digital signal and they are either 0 or 1. V_{out} could be computed for all the eight states of 3-bit digital signal. Network reduction method is used to obtain the expression for V_{out} for the 3-bit R-2R DAC.

$$V_{out} = -V_{ref}\left(\frac{D_2}{2} + \frac{D_1}{4} + \frac{D_0}{8}\right)$$

Example:
Let $V_{ref} = -5$ V and $D_2D_1D_0 = 011$

$$V_{out} = -(-5)\left(\frac{0}{2} + \frac{1}{4} + \frac{1}{8}\right) = 0.625 \text{ V}$$

Different values of V_{out} could be realized by changing the values of V_{ref} and the value of the feedback resistor (R_f). If $R_f = 2R$ for the 3-bit R-2R DAC, V_{out} would be 1.25 V. The general expression for n-bit R-2R DAC is:

$$V_{out} = -V_{ref} \sum_{i=1}^{n} \frac{D_{n-i}}{2^i}$$

References

1. Roberts P Achieving the best results with precision digital multi-meter measurements. Fluke Precision Measurement, USA
2. Gray N (2003) ABCs of ADCs, analog-to-digital converter basics. National Semiconductor Corporation, Santa Clara
3. Tutorial on sample-and-hold amplifiers, MT-090, Oct 2008. Analog Devices, USA
4. Understanding and minimizing ADC conversion errors. Application Note AN1636, STMicroelectronics, USA
5. Specifications and architectures of sample-and-hold amplifiers. Application Note AN775, National Semiconductor, July 1992
6. Renner K (2001) Analog reconstruction filter for HDTV using the THS8133, THS8134, THS8135, THS820. Application report, SLAA135, Sept 2001. Texas Instruments
7. Op-amp applications handbook (2005). Analog Devices
8. Kester W (2008) ADC architecture I: the flash converter. MT-20 Tutorial, Analog Devices, Oct 2008
9. Selecting A/D converters. Application Note AN016, Feb 1999. Renesas
10. Kester W (2008) ADC architectures III: sigma-delta basics. MT-022 Tutorial, Analog Devices, Oct 2008
11. Baker B (2011) How delta-sigma ADC work, Part 1. Texas Instruments, Analog Applications Journal
12. Razavi B (2016) The delta-sigma modulator. IEEE Solid State Magazine, Spring
13. Getting started with sigma-delta digital interface on applicable STM32 microcontrollers,. Application Note AN4990, STMicroelectronics, USA, Mar 2018

Chapter 12
Programmable Logic Devices

Abstract Programmable Logic Devices (PLDs) use combinational and sequential logic circuits as applicable to program logic functions. The operation of Read Only Memory (ROM), Mask ROM, Programmable ROM, Programmable Logic Array (PLA), Programmable Array Logic (PAL), Generic Array Logic (GAL), Complex Programmable Logic Device (CPLD) and Field Programmable Gate Array (FPGA) devices are presented. Applications of the devices are also presented.

12.1 Introduction

Basic gates, multiplexers and decoders are used for implementing logic functions and they have limitations in handling the number of variables. The limitations of multiplexers and decoders for implementing complex logic functions are overcome by using Programmable Logic Devices (PLDs). PLD integrated circuit blocks are available in various configurations. In general, PLDs contain an AND gate array followed by an OR gate array. Either AND array or OR array or both the arrays are programmable by users.

PLDs use combinational and sequential logic circuits as applicable to program logic functions. The operation of the following types of PLDs is presented:

(i) Programmed by IC manufacturers:

  – ROM (Read-Only Memory)
  – Mask ROM.

(ii) One-time user programmable logic devices:

  – PROM (Programmable Read-Only Memory)
  – PLA (Programmable Logic Array)
  – PAL (Programmable Array Logic).

(iii) User re-programmable logic devices:

  – GAL (Generic Array Logic)
  – CPLD (Complex Programmable Logic Device)

© Springer Nature Switzerland AG 2020   249
D. Natarajan, *Fundamentals of Digital Electronics*,
Lecture Notes in Electrical Engineering 623,
https://doi.org/10.1007/978-3-030-36196-9_12

– FPGA (Field Programmable Gate Array). One-time user programmable FPGA is also available.

12.2   Logic Devices Programmed by Manufacturers

Read-Only Memory (ROM) and Mask ROM devices are programmed by IC manufacturers as per the needs of customers. They are non-volatile memory devices and the stored program cannot be altered throughout the life of the devices.

12.2.1   ROM

ROMs are fabricated by placing discrete diodes in the form of arrays. After fabrication, each diode array circuit is equivalent to storing the bit, 1. The connections to selected diodes are removed for programming (storing) the bit, 0 as per customer codes. In 1965, Sylvania produced a 256-bit bipolar TTL ROM for Honeywell that was programmed one bit at a time by a skilled technician at the factory who physically scratched metal ink connections to selected diodes [1]. Scratching the metal ink is done to store the bit, 0 for realizing the codes of customers. ROMs with discrete diodes are obsolete and they are replaced by Mask ROMs.

12.2.2   Mask ROM

Mask ROMs (or simply ROMs) are programmed by the manufactures of the devices as per the binary codes of customers. Programming refers to defining the physical path of the logic signals from inputs to outputs in ROMs for realizing customer codes. Defining the paths of logic signals in ROMs is equivalent to storing binary data. The path of the logic signals as per customer codes are programmed (implemented) by manufacturers during the design of mask for the ROM ICs. Array of semiconductor devices such as Schottky diodes or transistors is integrated at mask level for programming. Mask level is the photolithography stage during the fabrication of ICs. The paths programmed at mask level cannot be altered by users or by turning power off to ROMs. Hence, ROMs are non-volatile memory devices. The combinations of the paths decide the memory capabilities of ROMs.

12.2.2.1 Capability of ROM

ROM is characterized by the number of input address lines, the number of output lines and bit size. In general, a ROM having n input address lines and m output lines is capable storing 2^n words having m number of bits provided the bit size (memory capacity) of the ROM is $2^n \times m$ bits. Two examples are provided for understanding the capabilities of ROM. Standard ROM ICs are available in various configurations.

Example-1:
Let the bit size of ROM to be 256 bits
Number of input address lines, n: 6
Number of output lines, m: 8
Number of 8-bit words that could be stored in the ROM $= 256/8 = 32$ i.e. 2^5.

Example-2:
Let the bit size of ROM to be 1024 bits
Number of input address lines, n: 8
Number of output lines, m: 4
Number of 4-bit words that could be stored in the ROM $= 1024/4 = 256$ i.e. 2^8.

12.2.3 Applications of ROM

ROM devices are used in computer systems and in many consumer products. Code conversion, character generations and logic function generation as per truth table are some of the applications of the devices. ROMs are also used in RFID applications for identification and retrieval of information [2]. Two applications of ROMs are presented.

12.2.3.1 Code Conversion

BCD 8421 codes could be converted into Excess-3 codes using a ROM device. ROM device for the code conversion consists of 4-to-10 decoder followed by an array of interconnecting diodes. The schematic diagram for the conversion is shown in Fig. 12.1. ABCD represents the BCD 8421 code and it is the input to the decoder. There are ten horizontal output lines from the decoder and they represent the ten words of Excess-3 code. There are four vertical output lines from ROM and they represent the four bits of Excess-3 code. The path between the output lines of the decoder and the output lines of the ROM are defined by an array of diodes in the figure. The outputs of ROM are the Excess-3 codes, $D_3 D_2 D_1 D_0$.

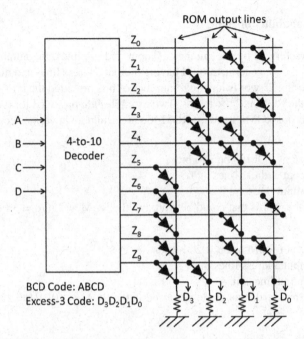

Fig. 12.1 BCD to Excess-3 code converter

If BCD input to the decoder, $ABCD = 0000$, the output line, Z_0, of the decoder goes High. As diode connections exist between the output line of the decoder and the output lines, D_1 and D_0, of ROM, the two output lines of the ROM go High. The other two output lines of the ROM, D_3 and D_2, remain Low. The Excess-3 code output, $D_3D_2D_1D_0$, is 0011.

If BCD input to the decoder, $ABCD = 0001$, the output line, Z_1, of the decoder goes High. Tracking the diode connections between the output line of the decoder and the output lines of ROM, the Excess-3 code output, $D_3D_2D_1D_0$, is 0100. The ten BCD codes and their conversion to Excess-3 codes are shown in Table 12.1.

12.2.3.2   Generating Logic Function

The BCD code inputs and Excess-3 code outputs for the MSB, D_3, in Table 12.1 are tabulated as the truth table of 4-variable function in Fig. 12.2. The logic function for D_3 in POS form is the logical sum of the implicants. The expression for the logic function is:

Table 12.1 BCD code inputs and Excess-3 code outputs of ROM

Decoder input: BCD code, ABCD	Decoder output	ROM output: Excess-3 code, $D_3 D_2 D_1 D_0$
0000	Z_0 goes High	0011
0001	Z_1 goes High	0100
0010	Z_2 goes High	0101
0011	Z_3 goes High	0110
0100	Z_4 goes High	0111
0101	Z_5 goes High	1000
0110	Z_6 goes High	1001
0111	Z_7 goes High	1010
1000	Z_8 goes High	1011
1001	Z_9 goes High	1100

A	B	C	D	D_3	
0	0	0	0	0	m_0
0	0	0	1	0	m_1
0	0	1	0	0	m_2
0	0	1	1	0	m_3
0	1	0	0	0	m_4
0	1	0	1	1	m_5
0	1	1	0	1	m_6
0	1	1	1	1	m_7
1	0	0	0	1	m_8
1	0	0	1	1	m_9

Fig. 12.2 Truth table of 4-variable logic function

$$D_3 = F(A, B, C, D) = \Sigma\, m\,(5, 6, 7, 8, 9)$$

The logic function, D_3, could be generated using the BCD to Excess code conversion circuit shown in Fig. 12.1. The function is available at the ROM output line, D_3, shown in the figure. The equivalent expression for the logic function is:

$$D_3 = F(A, B, C, D) = \Sigma\, m\,(5, 6, 7, 8, 9) = Z_5 + Z_6 + Z_7 + Z_8 + Z_9$$

Similarly, it could be observed in Fig. 12.1 that the logic functions generated at the ROM output lines, D_2, D_1 and D_0 are:

$$D_2 = F(A, B, C, D) = \Sigma\, m\,(1, 2, 3, 4, 9) = Z_1 + Z_2 + Z_3 + Z_4 + Z_9$$
$$D_1 = F(A, B, C, D) = \Sigma\, m\,(0, 3, 4, 7, 8) = Z_0 + Z_3 + Z_4 + Z_7 + Z_8$$

$$D_0 = F(A, B, C, D) = \Sigma\, m\,(0, 2, 4, 6, 8) = Z_0 + Z_2 + Z_4 + Z_6 + Z_8$$

12.2.3.3 Concept of AND and OR Arrays

BCD to Excess code conversion circuit shown in Fig. 12.1 is used to explain the concept of AND and OR arrays. The internal gate circuit of the 4-to-10 decoder consists of four inverters and ten AND gates. The variables, A, B, C and D, are the inputs to the inverters. The variables and the complemented outputs of the inverters are appropriately connected to the inputs of the AND gates. The set of interconnections to the AND gates is called AND array. There are ten sets of AND arrays in the decoder. The AND arrays are considered as fixed i.e. the interconnections between the inverters and AND gates of 4-to-10 decoder cannot be altered for programming.

There are four ROM output lines, D_3, D_2, D_1 and D_0, in Fig. 12.1. Each output line generates logic function in POS form. The output lines are considered as a set of OR arrays. There are four sets of OR arrays in the figure. The OR arrays of ROM are programmed (interconnected) by IC manufacturers using diodes for realizing logic functions. The OR arrays of ROM are considered as programmable. The concept of fixed and programmable arrays is used in PLDs.

12.3 Simple Programmable Logic Devices

ROMs are not economical for low volume manufacturing requirements. User programmable logic devices, PROM, PLA, PAL and GAL, are available for low volume requirements. The devices are grouped and termed as Simple Programmable Logic Devices (SPLDs).

SPLDs are classified as one-time user programmable and user re-programmable devices. PROM, PLA and PAL are one-time programmable devices. The stored programs in the SPLDs cannot be altered by users after programming.

GAL is user re-programmable SPLD. The stored program in the SPLD could be erased by users and new codes could be stored. The number of program/erase cycles is specified by the manufacturers of GAL devices. The construction and operation of the four SPLDs are presented.

12.3.1 Programmable Read-Only Memory

Programmable Read-Only Memory (PROM) devices are used in microprocessor based applications. PROM device consists of a set of input AND arrays and a set of output OR arrays. The AND array is fixed i.e. it is not programmable by users. The OR array is programmable by users as per the codes of the users. The codes cannot be altered after completing the programming process. Standard PROM ICs are available.

12.3.1.1 Representation of Arrays

The representations of fixed AND array and the programmable OR array are illustrated for a PROM device having two input variables for understanding. It is assumed that the 2-input PROM device stores four 3-bit words. The PROM has three output lines representing the bits of the words. The schematic diagram of the PROM device before programming is shown in Fig. 12.3a.

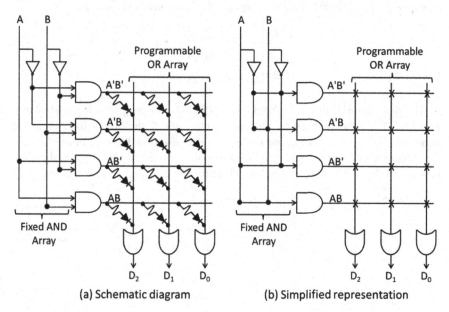

(a) Schematic diagram   (b) Simplified representation

Fig. 12.3 PROM with two inputs and three-outputs

It is cumbersome to represent AND and OR arrays as in Fig. 12.3a for PROM devices with higher capacity. The simplified representation of the arrays as shown in Fig. 12.3b is used. The AND array is simplified by having one input line for each AND gate. The interconnections between the input variables and the inputs of AND gates are shown by bullets in the figure. The bullets indicate that the AND array is fixed i.e. the array is not available for programming by users.

The OR array is simplified by showing crosses instead of series fuse-diode circuits. The crosses at interconnections indicate that OR array is available to users for programming.

12.3.1.2 Programming PROM

Initially, fuse-diode series circuits exist for all the logic paths of OR array as shown by the crosses in Fig. 12.3b. Crosses indicate that the bit, 1, is stored in all the logic paths of PROM. The OR array is programmed by users. Programming means storing the bit, 0 in the required logic paths of the PROM. It is performed by blowing off the fuses in the required paths. Appropriate burn-in stimulus controller is used for programming the OR array. The controller is capable of passing high current through selected fuses. The fuses that are not required as per the needs of codes are selected and burnt by passing higher current through the fuses. The programmed logic functions are available at the output OR gates (D_2, D_1 and D_0) of PROM. For example, assume that the required output logic functions of PROM are:

$$F(D_2) = m(2, 3)$$
$$F(D_1) = m(0, 2)$$
$$F(D_0) = m(1, 2, 3)$$

The realization of the three functions in PROM after programming is shown in Fig. 12.4. $F(D_2)$ is realized by blowing of the fuses in the output line, D_2 at the logic path intersections at m_0 and m_1. $F(D_1)$ is realized by blowing of the fuses in the output line, D_1 at the logic path intersections at m_1 and m_3. $F(D_0)$ is realized by blowing of the fuses in the output line, D_0 at the logic path intersection at m_0.

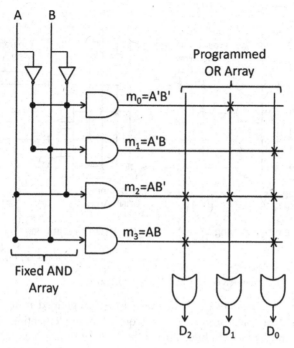

Fig. 12.4 Representation of PROM after programming

12.3.2 Programmable Logic Array

Programmable Logic Array (PLA) devices use PROM technology for programming logic functions. The representation of PLA device with 2-input AND array and 2-output OR array before programming is shown in Fig. 12.5a. All the paths of AND and OR arrays are shown with crosses, indicating that both the arrays are user programmable. The arrays have series fuse-diode circuits. The required logic paths of the arrays are programmed to 0 by blowing off the appropriate fuses for implementing logic functions.

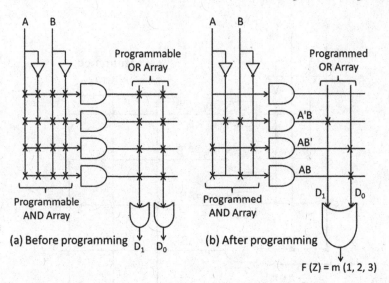

Fig. 12.5 PLA with 2-input AND array and 2-output OR array

Assume that F (Z) = m (1, 2, 3) needs to be programed in the PLA device shown in Fig. 12.5a. The fuses that are not required for implementing the logic function are blown off by passing high current through the fuses. The representation of programmed PLA for implementing the logic function, F (Z), is shown in Fig. 12.5b.

12.3.3 *Programmable Array Logic*

PAL is the trade mark of Advanced Micro Devices, USA. PAL devices use PROM technology for programming logic functions. The devices have user programmable input AND array and fixed output OR array. The AND array has series fuse-diode circuits. The representation of PLA device with 2-input programmable AND array and 2-output fixed OR array is shown in Fig. 12.6a. The paths of AND array are shown with cross interconnections and that of OR arrays are shown with bullet interconnections. Logic functions are programmed by blowing off the appropriate fuses in the AND array.

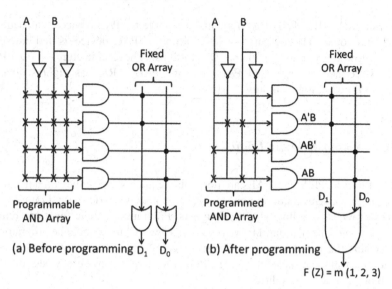

Fig. 12.6 Representation of PAL with two inputs and outputs

Assume that $F(Z) = m(1, 2, 3)$ needs to be programed in the PAL device shown in Fig. 12.6a. The fuses that are not required for programming the logic function are blown off by passing high current through the fuses. The representation of programmed PAL for obtaining the logic function, $F(Z)$, is shown in Fig. 12.6b.

12.3.3.1 PAL Architectures

The architecture of a PAL device is defined by the output logic of the device. The simplified schematic diagram of PAL in Fig. 12.7 shows combinational OR output logic. The architecture of the PAL device is termed as combinational PAL. Standard ICs (Ex.: PAL16L8) are available for the combinational PAL architecture.

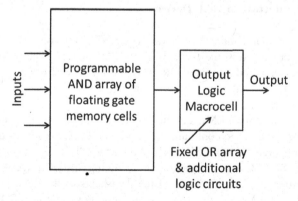

Fig. 12.7 General representation of GAL device

Standard ICs (Ex.: PAL 16R8) are also available for PAL devices with sequential output logic. The outputs of AND array of PAL device is fed into OR array followed by registers using D flip-flops. The architecture of the PAL device is termed as registered PAL. The IC, PAL 16R8, has eight sequential (registered) output logic circuits.

12.3.4 Generic Array Logic

ROMs, PROMs, PLAs and PALs are non-volatile memories but they do not support debugging requirements during product development. Generic Array Logic (GAL) devices provide more flexibility to designers. Programmed logic functions remain permanent in the devices until they are erased. New functions could be programmed after erasing the stored functions. The re-programmable feature in GAL devices is realized by using Erasable PROM (EPROM) technology in the devices. EPROM technology is briefly introduced.

12.3.4.1 EPROM Technology

Erasable PROMs use an array of floating gate MOSFET memory cells for storing bits. Each memory cell is capable of storing one bit of data. Initially, all the memory cells are in conducting state and it is equivalent to storing the bit, 1. The memory cells in conducting state are in erased state. Selected memory cells are then programmed to the bit, 0 by forcing the memory cells to non-conducting state. Programmed bits could be erased electrically or by using ultra-violet rays. GAL devices use electrically erasable floating gate MOSFET memory cells for programming. The structure and logic levels of memory cell are explained in Appendix.

12.3.4.2 Architecture of GAL Device

GAL has programmable input AND array. The AND array consists of floating gate memory cells organized in the forms of rows and columns. The outputs of AND array of the device is fed into fixed OR array followed by additional logic circuits. The functional block containing OR array and the additional logic circuits is called Output Logic Macrocell (OLMC).

GAL devices contain OLMC circuits in many configurations. The architecture of GAL allows the users to configure the required OLMC circuit for their applications. The configurations of OLMC include the two PAL architectures explained in Sect. 12.3.3.1. GAL devices are capable of emulating many PAL devices. Standard ICs are available for GAL devices and the datasheets of the devices indicate the configurations of OLMC circuits for programming. For example, GAL16V8 could be configured with registered or Complex or simple OLMC circuits. The general representation of GAL device is shown in Fig. 12.7.

12.4 Complex Programmable Logic Device

The number of product terms of logic function increases substantially with the ever increasing integration of functional requirements with product design. SPLDs are not suitable for programming logic functions with large number of product terms. Complex Programmable Logic Devices (CPLDs) are available for programming hundreds of product terms. CPLDs use electrically erasable technology. The devices are in-system programmable using software after surface mounting on printed circuit boards (PCBs). The devices are re-programmable by users. Re-programming could be done without de-soldering the devices from PCBs.

CPLDs are available from many manufacturers. Although the basic function of CPLDs of all manufacturers remains same, the features and the architectures of the devices vary among the manufacturers. The basic architecture of CPLD is presented providing the overview of the device operation. It is based on the architecture of Lattice Semiconductor Corporation's Mach 1 and Mach 2 CPLD families. Mastering the operation of CPLD requires understanding the device datasheets of manufacturers and designing logic circuits using a suitable software package. Learning efforts should also be supported by experimenting with devices for developing logic function.

12.4.1 Basic Architecture

The simplified architecture of CPLD is shown in Fig. 12.8. CPLD consists of multiple programmable logic devices (PLDs). The PLDs are called Logic Array Blocks (LABs) or Programmable Array Logic (PAL) blocks. The number of LABs in CPLD is specified in the datasheets of device manufacturers. Four LABs are shown in the figure.

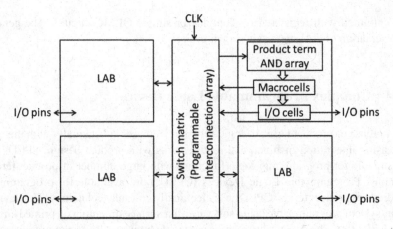

Fig. 12.8 Simplified representation of the architecture of CPLD

Logic Array Blocks (LABs) are fabricated on a single chip and the LABs are interconnected by a programmable switch structure. The switch structure is called switch matrix or Programmable Interconnect Array (PIA). User inputs to CPLD are routed through switch matrix to LABs. The outputs of LABs are the outputs of CPLD and they are also routed through the switch matrix. User inputs to CPLD are the requirements of logic circuits for new designs. The output of CPLD is the simplified logic function.

12.4.1.1   Functions of Logic Array Block

Each Logic Array Block (LAB) consists of product-term AND array, macrocells and I/O (input and output) cells. The number of macrocells per LAB is specified in the datasheets of device manufacturers. Product-term AND array converts user inputs into product terms. The product terms are grouped and allocated to macrocells automatically by software. Macrocells are capable of generating registered or combinational outputs. The I/O cells of LAB use tristate output buffer. The buffer could be enabled to provide the output of CPLD or disabled for providing inputs to the device.

12.5   Field Programmable Gate Array

Field Programmable Gate Array (FPGA) has higher logic density than CPLD. CPLD has thousands of gates and FPGA has tens of thousands of gates. Two types of FPGAs are available. Re-programmable devices using Static Random Access Memory (SRAM) technology and one-time programmable devices using anti-fuse technology are available.

SRAM technology is volatile i.e. the configuration data of FPGA is lost without power source. An external memory device such as Flash memory chip is required to support the functioning of FPGA. The configuration data of FPGA is stored in the external memory device. FPGA reads the data from the memory device when the device is powered on. FPGA devices using SRAM technology are re-programmable and they are popular for many applications. SRAM based FPGA is used to design laboratory experiments illustrating simple digital signal processing applications.

Electrical path exists in fuses and the path becomes open with excessive current. Anti-fuse is the opposite of fuses. Electrical path is in open state normally using a dielectric material. High voltage is used to force a current through the dielectric, destroying the dielectric material. The electrical path becomes conductive. FPGA devices that use anti-fuse technology are one-time programmable.

For low to medium volume productions, FPGAs provide cheaper solution and faster time to market as compared to Application Specific Integrated Circuits (ASICs) which normally require a lot of resources in terms of time and money to obtain first device [3]. The output of FPGA could be combinational or registered type using flip-flops. FPGA architecture with register-intensive output is best suited for many logic design applications [4].

12.5.1 *General Architecture*

FPGA has Configurable Logic Blocks (CLBs) and I/O cells, interconnected by Programmable Wiring Channels (PWCs). Logic functions are programmed in FPGA using CLBs and PWCs. The programmed logic functions are stored in the SRAM logic cells or in the logic cells created by using anti-fuse technology. The logic cells are the memory cells and they are distributed throughout FPGA.

The simplified general architecture of FPGA with four CLBs with PWCs and I/O cells is shown in Fig. 12.9 although standard FPGA ICs contain hundreds of CLBs. The I/O cells are tristate output buffer circuits. The buffer could be enabled to provide the output of FPGA or disabled for providing inputs to the device.

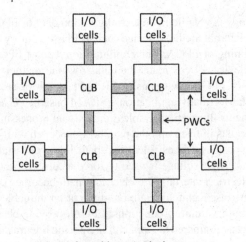

CLB: Configurable Logic Block
PWCs: Programmable Wiring Channels
I/O cells: Input/Output cells

Fig. 12.9 Simplified general architecture of FPGA

12.5.1.1   Configurable Logic Block

Each CLB contains several programmable logic blocks and PWCs. Each logic block contains combinational and sequential hardware. Look-Up Tables (LUTs) for Combinational hardware and flip-flops are the components of logic blocks. Example of LUT is presented in the subsequent section. CLBs are capable of implementing a wide range of logic functions for digital signal processing applications.

PWCs are programmable switches. The switches are programmed by users to interconnect CLBs and the components of logic blocks within CLBs for implementing logic functions. The configured data of CLBs and PWCs is stored in the memory cells of FPGA.

12.5.1.2   Example of LUT with SRAM Cells

LUT defines the outputs for the set of inputs of a digital hardware. LUT with k inputs could be programmed to implement any Boolean function with k variables and it requires 2^k SRAM bits. Truth table with three input variables is shown in Fig. 12.10a. Implementing the logic function using LUT is shown in Fig. 12.10b. LUT requires 2^3 i.e. eight SRAM cells and the cells are programmed to store the data inputs for implementing the logic function using 8:1 MUX. The SRAM cells and the stored data inputs in the cells are shown in the figure.

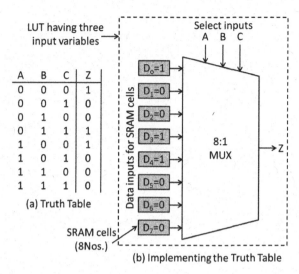

Fig. 12.10 Implementing logic function using LUT

References

1. Finn CL (1972) All semiconductor memory system includes read-only and read/write chips. HP J 22–24
2. Ahsan K, Shah H, Kingston P (2010) RFID applications: an introductory and exploratory study. IJCSI Int J Comput Sci Issues 7(1):3
3. Farooq U et al (2012) Tree-based heterogeneous FPGA architectures. Springer
4. CPLDs versus FPGAs, comparing high-capacity programmable logic, product information bulletin, vol 18. Altera Corporation, Feb 1995

Chapter 13
Design of Sequential Logic Circuits

Abstract Finite State Machine (FSM) is a tool for designing sequential logic circuits by defining states. The general models of Moore and Mealy machines are presented. The design of Sequence detector using Moore and Mealy machines are explained and illustrated with an example. The preparation of algorithmic state machine chart and the application of state reduction techniques (Row elimination and Implication table methods) are also illustrated with examples.

13.1 FSM Models

Finite State Machine (FSM) is a tool for designing sequential logic circuits by defining states. The number of states for designing the logic circuits is defined by users. Algorithms describing the operational sequences of sequential logic circuits are developed for accomplishing the design of the circuits. The algorithms are simulated using FSM modelling tools for verifying the design of sequential circuits before hardware realization.

FSM designs are used in a variety of applications covering a broad range of performance and complexity; low-level controls of microprocessor-to-VLSI-peripheral devices, bus arbitration in conventional microprocessors, custom bit-slice microprocessors, data encryption and decryption, and transmission protocols are but a few examples [1]. Moore and Mealy modelling tools are available for the design of sequential logic circuits. In simpler terms, they are called as Moore machine and Mealy machine.

13.1.1 General Models of Moore and Mealy Machines

The general models of Moore and Mealy machines are shown in Fig. 13.1. The input to the state machines is denoted by X. The output of the machines is denoted by Y. The state machines consist of three functional blocks. Input combinational block, registers and output combinational block are the three blocks. The input combinational logic

© Springer Nature Switzerland AG 2020
D. Natarajan, *Fundamentals of Digital Electronics*,
Lecture Notes in Electrical Engineering 623,
https://doi.org/10.1007/978-3-030-36196-9_13

circuit block generates next input state to the flip-flops of the register block. The register block stores the present output states of the flip-flops. The present output state is fed back to the input combinational logic block to generate the next output state for the register. The register block is normally modelled as edge-triggered D flip-flops [2]. The output combinational logic block delivers the final output, Y.

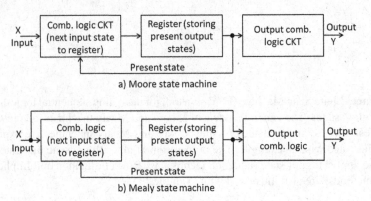

Fig. 13.1 Moore and Mealy state machines

It could be seen from Fig. 13.1a that the output (Y) of Moore state machine depends only on the present state of the register. The output (Y) of Mealy state machine in Fig. 13.1b depends on both the present state of the register and the input, X.

Both the state machines are used for the design of sequential logic circuits. Generally, sequential logic design using Mealy machine requires less hardware for implementation compared to design using Moore machine. The operation of the circuit is also relatively faster. Mealy machine is susceptible to glitches in the input signal and passes the undesired disturbances to the output; Moore machine is not susceptible to glitches and output logic glitches could be overcome by buffering schemes [3].

13.1.2 Designing Sequential Logic Circuits

The general approach for designing synchronous sequential logic circuits using Moore and Mealy machines is:

– Prepare design requirements
– Represent the requirements in the form of state transition diagram
– Prepare next state table
– Decide the number of flip-flops
– Encode the states
– Prepare final state table with encoding

– Obtain logic functions for the combinational circuit blocks from the final state table using K-maps
– Implement the sequential logic circuit.

The approach is illustrated for designing binary sequence detector circuit using Moore and Mealy machines. In baseband digital communication channels, binary sequence detectors are used at the receiving end to detect defined bit sequences to identify the beginning and ending of message [4].

13.2 Sequence Detector Using Moore Machine

Assume that a sequential logic circuit should be designed for detecting the binary sequence, 110, in a binary stream using Moore machine. The binary stream is the input, X, to the Moore machine. One bit of the serial binary stream is applied as input for every clock cycle. The output, Y, goes High when the sequence, 110, is detected; otherwise, Y remains Low. The sample binary stream to the (X) input of Moore machine and the output (Y) binary stream are shown in Fig. 13.2. The output (Y) is 1 when the sequence, 110, is detected in the input binary stream. The sequence, 110, in the input binary stream and the High output in output binary stream are shown in bold in the figure.

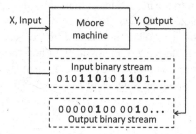

Fig. 13.2 Input binary stream and output of Moore machine

13.2.1 State Transition Diagram

The state transition diagram for detecting the sequence, 110, is prepared in two stages. Stage-1 state transition diagram is prepared assuming straight occurrence of the sequence, 110. Stage-2 state transition diagram is prepared by adding overlapping occurrences in the state-1 transition diagram. Stage-2 diagram represents the final state transition diagram of the Moore machine for detecting the sequence, 110.

13.2.1.1 Stage-1: Straight Occurrences

Stage-1 state transition diagram is shown in Fig. 13.3. Let the initial state of Moore machine for detecting the sequence, 110, be A. The state, A, is reset to 0. State-A is shown by a circle in the figure. The output, Y, is 0 and it is indicated within the circle.

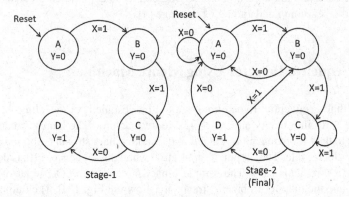

Fig. 13.3 Moore state transition diagram for detecting the sequence, 110

Straight occurrence of the sequence, 110, means that the bits, 1, 1 and 0 are received in order as input (X). When the first bit i.e. X = 1 is received, the Moore machine changes from A to the next state, B. State-B is also shown by a circle. States A and B are linked by a curved arrow. The received bit i.e. X = 1 is indicated on the curved arrow. Y remains at 0. When the second bit i.e. X = 1 is received, State-B changes to changes to State-C. Y remains at 0. When the third bit i.e. X = 0 is received, State-C changes to State-D. The output, Y, changes to 1. The state changes, inputs and outputs are shown as Stage-1 in Fig. 13.3. The sequence, A-B-C-D, is straight occurrence.

13.2.1.2 Stage-2: Overlapping Occurrences

Stage-2 state transition diagram in Fig. 13.3 shows the state changes in each state when a bit other than the bit required for the sequence, 110, is received. Such state changes are overlapping occurrences. Overlapping occurrence might result in remaining in the same state or changing to another state. Remaining in the same state is shown by small arc around the circle. Changing to another state is shown by appropriate arrows.

Overlapping occurrences are identified considering the possibility of receiving the sequence, 110, subsequently. Two examples of overlapping occurrences are provided. Assume that bit, 1 (X = 1), is received in State-C. The state remains in State-C until bit, 0, is received. Y also remains at 0. If bit, 0, is received in State-C, the state changes to State-D. The output, Y, becomes 1. The sequence, B-C-C-D, is one of the overlapping occurrences. Similarly, the sequence, D-B-C-D is another overlapping occurrence, when bit, 1, is received in State-D. The completed stage-2 state transition diagram is the final state transition diagram for the detecting the sequence, 110.

13.2.2 Next State Table

Next state table is prepared from state transition diagram. The state table is prepared using the symbolic identifications (A, B, C and D) of states in transition diagram. The preparation of next state table for detecting the sequence, 110 is straight forward. Consider the present state, A. If $X = 0$ in State-A, the state does not change i.e. the next state is A. The output, Y, remains at 0. If $X = 1$ in State-A, the state changes to B i.e. the next state is B. Y continue to remain at 0. Similarly, the next states for each present state considering the input, X, are obtained. The completed next state table for detecting the sequence, 110, is shown in Fig. 13.4.

Present state	Input, X	Next state	Output, Y
A	0	A	0
A	1	B	0
B	0	A	0
B	1	C	0
C	0	D	1
C	1	C	0
D	0	A	0
D	1	B	0

Fig. 13.4 Moore next state table for detecting the sequence, 110

13.2.3 Number of Flip-Flops

The number of flip-flops is decided for the design of sequential circuit. Two flip-flops could represent four states i.e. 2^2 states; three flip-flops could represent 8 states i.e. 2^3 states and so on. Two flip-flops are adequate to represent the four states for detecting the sequence, 110. The flip-flops are identified as FF-1 and FF-2. D_0 and Q_0 are the input and output of FF-1. D_1 and Q_1 are the input and output of FF-2.

13.2.4 Encoding of States

The symbolic identifications (A, B, C and D) of states are encoded for obtaining the logic function of sequential circuit. Many schemes are available for encoding the symbolic identifications of the states. Binary, Gray code and one-hot encoding schemes are commonly used state assignment schemes and obtaining optimal assignment is very difficult [3].

In general, if state transition diagram has four states, the binary scheme assignments to the states are 00, 01, 10 and 11. If state transition diagram has three states, the assignments to the states are 00, 01 and 10. In the Gray code scheme, if state transition diagram has four states, the assignments to the states are 00, 01, 11 and 10. If state transition diagram has three states, the Gray code assignments to the states are 00, 01 and 11. The design of sequential circuit for detecting the sequence, 110, is presented with binary and Gray code encoding.

13.2.5   Final State Table with Binary Encoding

Two flip-flops are used in the design of sequential circuit for detecting the sequence, 110. Present and next states are identified by the outputs of the two flip-flops. The present states of the flip-flops are identified by Q_1 and Q_0. Two bits are used to encode the symbolic identification of the states. The 2-bit binary assignments to the symbolic identification of the present states in preparing final state table are:

- Q_1 and Q_0 for State-A: 00
- Q_1 and Q_0 for State-B: 01
- Q_1 and Q_0 for State-C: 10
- Q_1 and Q_0 for State-D: 11.

13.2.5.1   Final State Table

The next state table in Fig. 13.4 is modified to prepare final state table. Q_1 and Q_0 are indicated for present state. Q_1^+ and Q_0^+ are indicated as next state. The binary codes of the symbolic identifications of states are indicated in the table. Two more columns are added showing the inputs (D_1 and D_2) of the two flip-flops. The final state table is shown in Fig. 13.5.

Present state		Input, X	Next state		Output, Y	D_1	D_0
Q_1	Q_0		Q_1^+	Q_0^+			
0	0	0	0	0	0	0	0
0	0	1	0	1	0	0	1
0	1	0	0	0	0	0	0
0	1	1	1	0	0	1	0
1	0	0	1	1	1	1	1
1	0	1	1	0	0	1	0
1	1	0	0	0	0	0	0
1	1	1	0	1	0	0	1

Fig. 13.5   Moore final state table with binary assignment

The states of D_1 are identified by examining Q_1^+ and Q_1. If Q_1^+ is 1, D_1 is 1 in D flip-flop; Q_1 is don't care input. Accordingly, the states of D_1 are indicated in the final state table. Similarly, the states of D_0 are identified by examining Q_0^+ and Q_0. Accordingly, the states of D_0 are indicated in the final state table.

13.2.6 Logic Functions with Binary Encoding

The simplified logic functions for the combinational logic circuit driving the flip-flops are obtained using K-maps. The input (X), present inputs (Q_1 and Q_0), flip-flop inputs (D_1 and D_0) and output (Y) of the final state table are used for obtaining the logic functions. The final state table data is split into three tables, namely, D_1 table, D_0 table and Y table. The three tables are shown in Fig. 13.6a–c. The K-maps for the tables and the simplified logic functions for D_1, D_0 and Y are also shown in the figure.

Q_1	Q_0	X	D_1
0	0	0	0
0	0	1	0
0	1	0	0
0	1	1	1
1	0	0	1
1	0	1	1
1	1	0	0
1	1	1	0

Q_1	Q_0	X	D_0
0	0	0	0
0	0	1	1
0	1	0	0
0	1	1	0
1	0	0	1
1	0	1	0
1	1	0	0
1	1	1	1

Q_1	Q_0	Y
0	0	0
0	1	0
1	0	1
1	1	0

	X'	X
$Q_1'Q_0'$	0	0
$Q_1'Q_0$	0	1
Q_1Q_0	0	0
Q_1Q_0'	(1	1)

	X'	X
$Q_1'Q_0'$	0	1
$Q_1'Q_0$	0	0
Q_1Q_0	0	1
Q_1Q_0'	1	0

	Q_0'	Q_0
Q_1'	0	0
Q_1	1	0

$Y = Q_1Q_0'$

(c) Y Table/ K-map

$D_1 = X Q_1'Q_0 + Q_1Q_0'$

$D_0 = X Q_1'Q_0' + XQ_1Q_0 + X'Q_1Q_0'$

$D_0 = X (Q_1' \odot Q_0) + X'Q_1Q_0'$

(a) D_1 Table/ K-map   (b) D_0 Table/ K-map

Fig. 13.6 Moore K-maps with binary assignment

13.2.7 Sequential Logic Circuit with Binary Encoding

The logic functions for the Moore sequence detector with binary assignment for the sequence, 110, are implemented. The schematic diagram of the sequence detector is shown in Fig. 13.7. XNOR gate could also be used for D_0.

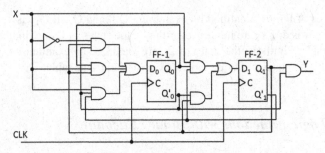

Fig. 13.7 Moore sequence detector with binary assignment

13.2.8   Sequential Logic Circuit with Gray Code Encoding

The final state table for the sequence detector for the sequence, 110 with Gray code assignments is shown in Fig. 13.8. The Gray code assignments to the states are:

Present input		Input, X	Next state		Output, Y	D_1	D_0
Q_1	Q_0		Q_1^+	Q_0^+			
0	0	0	0	0	0	0	0
0	0	1	0	1	0	0	1
0	1	0	0	0	0	0	0
0	1	1	1	1	0	1	1
1	1	0	1	0	0	1	0
1	1	1	1	1	0	1	1
1	0	0	0	0	1	0	0
1	0	1	0	1	0	0	1

Fig. 13.8 Moore final state table with Gray code assignment

- Q_1 and Q_0 for State-A: 00
- Q_1 and Q_0 for State-B: 01
- Q_1 and Q_0 for State-C: 11
- Q_1 and Q_0 for State-D: 10.

13.2.8.1   Obtaining Logic Functions

The three K-maps for the final state table are also shown in Fig. 13.9. The simplified logic functions for D_1, D_0 and Y are also shown in the figure.

Q_1	Q_0	X	D_1
0	0	0	0
0	0	1	0
0	1	0	0
0	1	1	1
1	1	0	1
1	1	1	1
1	0	0	0
1	0	1	0

Q_1	Q_0	X	D_0
0	0	0	0
0	0	1	1
0	1	0	0
0	1	1	1
1	1	0	0
1	1	1	1
1	0	0	0
1	0	1	1

Q_1	Q_0	Y
0	0	0
0	1	0
1	0	1
1	1	0

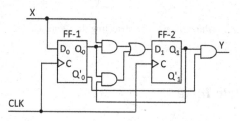

$D_1 = XQ_0 + Q_1Q_0$

$D_0 = X$

$Y = Q_1Q_0'$

(a) D_1 Table/ K-map   (b) D_0 Table/ K-map   (c) Y Table/ K-map

Fig. 13.9 Moore K-maps with Gray code assignment

13.2.8.2   Implementation

The logic functions for the Moore sequence detector with Gray code assignment for the sequence, 110, are implemented. The schematic diagram of the sequence detector is shown in Fig. 13.10. The logic circuit with Gray code assignment is simpler than with binary assignment for the states.

Fig. 13.10 Moore sequence detector with Gray code assignment

13.3   Sequence Detector Using Mealy Machine

The design of Mealy sequence detector is illustrated for detecting the binary sequence, 110, in a binary stream. The general procedure for the design of Mealy sequence detector is same as that used for Moore sequence detector.

13.3.1 State Transition Diagram

State transition diagram for detecting the sequence, 110, using Mealy machine is shown in Fig. 13.11. Both straight and overlapping occurrences are shown in the figure. The input bit to the sequence detector is X and the output bit of the detector is Y. Let A be the state before applying the input binary stream to the sequential logic circuit. The state, A, is reset to 0. If the sequence detector receives bit 1 i.e. X = 1, the state, A changes to state, B. The output, Y is 0. The input and output are indicated on the curved arrow linking the two states in Fig. 13.11. Instead, if X = 1, Y = 0 and the state remains at A. The input and output are shown by an arc around the state, A. The procedure is continued for the states, B and C for the sequence, 110. It could be observed that Mealy sequence detector has only three states.

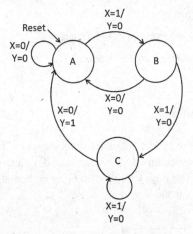

Fig. 13.11 Mealy state transition diagram for detecting the sequence, 110

13.3.2 Next State Table

The next state table for the Mealy sequence detector for the sequence, 110 is shown in Fig. 13.12. The table is prepared using state transition diagram.

Present state	Input, X	Next state	Output, Y
A	0	A	0
A	1	B	0
B	0	A	0
B	1	C	0
C	0	A	1
C	1	C	0

Fig. 13.12 Mealy next state table for detecting the sequence, 110

13.3.3 *Sequential Logic Circuit with Binary Encoding*

Final state table with binary assignment to states, obtaining logic functions and the implementation of sequential logic circuit for the sequence detector are presented.

13.3.3.1 Final State Table

Two flip-flops are required for representing three states. The final state table with binary assignments to the states is shown in Fig. 13.13. The inputs to the flip-flops are also shown in the table.

Present input		Input, X	Next state		Output, Y	D_1	D_0
Q_1	Q_0		Q_1^+	Q_0^+			
0	0	0	0	0	0	0	0
0	0	1	0	1	0	0	1
0	1	0	0	0	0	0	0
0	1	1	1	0	0	1	0
1	0	0	0	0	1	0	0
1	0	1	1	0	0	1	0

Fig. 13.13 Mealy final state table with binary assignment

13.3.3.2 Obtaining Logic Functions

The final state table data is split into three tables. The K-maps for the tables are also shown in Fig. 13.14. The simplified logic functions for D_1, D_0 and Y are obtained from the K-maps and they are also shown in the figure.

Q_1	Q_0	X	D_1
0	0	0	0
0	0	1	0
0	1	0	0
0	1	1	1
1	0	0	0
1	0	1	1

	X'	X
$Q_1'Q_0'$	0	0
$Q_1'Q_0$	0	1
Q_1Q_0	X	X
Q_1Q_0'	0	1

$D_1 = XQ_0 + XQ_1$

(a) D_1 Table/ K-map

Q_1	Q_0	X	D_0
0	0	0	0
0	0	1	1
0	1	0	0
0	1	1	0
1	0	0	0
1	0	1	0

	X'	X
$Q_1'Q_0'$	0	1
$Q_1'Q_0$	0	0
Q_1Q_0	X	X
Q_1Q_0'	0	0

$D_0 = X\,Q_1'Q_0'$

(b) D_0 Table/ K-map

Q_1	Q_0	X	Y
0	0	0	0
0	0	1	0
0	1	0	0
0	1	1	0
1	0	0	1
1	0	1	0

	X'	X
$Q_1'Q_0'$	0	0
$Q_1'Q_0$	0	0
Q_1Q_0	X	X
Q_1Q_0'	1	0

$Y = X'Q_1$

(c) Y Table/ K-map

Fig. 13.14 Mealy K-maps with binary assignment

13.3.3.3   Implementation

The logic functions for the Mealy sequence detector with binary assignments for the sequence, 110, are implemented. The schematic diagram of the sequence detector is shown in Fig. 13.15.

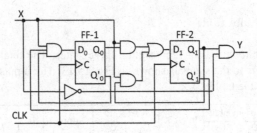

Fig. 13.15   Mealy sequence detector with binary assignment

13.3.4   Sequential Logic Circuit with Gray Code Encoding

The final state table with Gray code assignment to the states is shown in Fig. 13.16. The inputs of D flip-flops are also shown in the figure.

Present input		Input, X	Next state		Output, Y	D_1	D_0
Q_1	Q_0		Q_1^+	Q_0^+			
0	0	0	0	0	0	0	0
0	0	1	0	1	0	0	1
0	1	0	0	0	0	0	0
0	1	1	1	1	0	1	1
1	1	0	0	0	1	0	0
1	1	1	1	1	0	1	1

Fig. 13.16   Mealy final state table with Gray code assignment

13.3.4.1   Obtaining Logic Functions

The final state table data is split into three tables. The K-maps for the tables are also shown in Fig. 13.17. The simplified logic functions for D_1, D_0 and Y are obtained from the K-maps and they are also shown in the figure.

Q_1	Q_0	X	D_1
0	0	0	0
0	0	1	0
0	1	0	0
0	1	1	1
1	1	0	0
1	1	1	1

Q_1	Q_0	X	D_0
0	0	0	0
0	0	1	1
0	1	0	0
0	1	1	1
1	1	0	0
1	1	1	1

Q_1	Q_0	X	Y
0	0	0	0
0	0	1	0
0	1	0	0
0	1	1	0
1	1	0	1
1	1	1	0

	X'	X
$Q_1'Q_0'$	0	0
$Q_1'Q_0$	0	1
Q_1Q_0	0	1
Q_1Q_0'	X	X

$D_1 = XQ_0$

	X'	X
$Q_1'Q_0'$	0	1
$Q_1'Q_0$	0	1
Q_1Q_0	0	1
Q_1Q_0'	X	X

$D_0 = X$

	X'	X
$Q_1'Q_0'$	0	0
$Q_1'Q_0$	0	0
Q_1Q_0	1	0
Q_1Q_0'	X	X

$Y = X'Q_1$

(a) D_1 Table/ K-map   (b) D_0 Table/ K-map   (c) Y Table/ K-map

Fig. 13.17 Mealy K-maps with Gray code assignment

13.3.4.2 Implementation

The logic functions for the Mealy sequence detector with Gray code assignments for the sequence, 110, are implemented. The schematic diagram of the sequence detector is shown in Fig. 13.18.

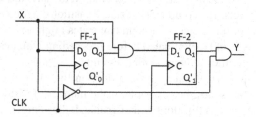

Fig. 13.18 Mealy sequence detector with Gray code assignment

13.4 Algorithmic State Machine Chart

Algorithmic state machine (ASM) chart represents the states of FSM using flow chart symbols. ASM chart is an alternative to state transition diagram of FSM. The chart uses state box, decision box and output box. The number of boxes is decided by the requirements of sequential logic design.

A state is identified by a rectangle with its symbolic representation indicated on the top of the rectangle. The decision box tests the input bit, X. The output box indicates the output, Y. State changes are indicated by the exit paths from decision box. The ASM chart for the Mealy sequence detector for the sequence, 110 is shown in Fig. 13.19.

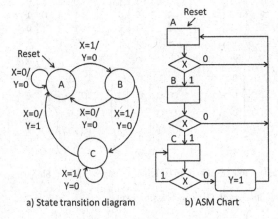

a) State transition diagram   b) ASM Chart

Fig. 13.19 ASM chart of Mealy sequence detector for the sequence, 110

13.5   State Reduction

Developing state transition diagram for system requirements is the first step to design sequential logic circuits. The complexity of state transition diagram increases with the complexity of system requirements. Redundant states might exist in state transition diagram and they should be eliminated before proceeding with design. Eliminating redundant states simplifies sequential logic circuit.

Generally, the number of gates and flip-flops in the logic circuit reduces after eliminating redundant states. For example, if the number of states is reduced from twelve to ten, the number of flip-flops remains same. If the number of states is reduced from twelve to eight or less, the number of flip-flops reduces. Don't care conditions are also effectively used for obtaining simplified logic functions after eliminating redundant states.

Next state table is used for identifying redundant states. Two states of next state table are redundant if they change to the same or equivalent states for every input (X = 0 and X = 1) and generate the same output (Y) for both X = 0 and X = 1. Row elimination and Implication table methods are used to identify redundant states. The methods are illustrated with hypothetical next state tables having seven states.

13.5.1 Row Elimination Method

A hypothetical next state table with seven states is shown in Fig. 13.20. The table is used for eliminating redundant states using Row elimination method. Each state is examined with other states to identify redundant states and eliminate them.

Present	Next state		Output, Y	
state	X=0	X=1	X=0	X=1
A	A	B	0	0
B	B	D	0	1
C	D	G	0	0
D	D	B	1	1
E	B	D	0	1
F	F	E	1	1
G	C	A	0	0

Fig. 13.20 Next state table for state reduction

13.5.1.1 Eliminating Redundant States

Consider the states, A and B. The outputs (Y) at $X = 1$ for the states differ. Hence, A and B are not redundant. Examining the states, A and C, the outputs when $X = 0$ and $X = 1$ are same; but A changes to D at $X = 0$ and B changes to G. Hence, A and D are not redundant. The process is continued by examining A-D, A-E, A-F, A-G, B-C, B-D and so on until the states, F and G are examined.

Consider the states, B and E. When $X = 0$, B remains at B and it changes to D at $X = 1$. When $X = 0$, E changes to the same state, B and it changes to the same state, D at $X = 1$. In other words, B and E change to the same states for the inputs. The outputs when $X = 0$ and $X = 1$ are also same. Hence, B and E are redundant states. State-E is eliminated. The reduced next state table is shown in Fig. 13.21.

Present	Next state		Output, Y	
state	X=0	X=1	X=0	X=1
A	A	B	0	0
B	B	D	0	1
C	D	G	0	0
D	D	B	1	1
F	F	E	1	1
G	C	A	0	0

Fig. 13.21 Reduced next state table using Row elimination method

13.5.1.2 Limitations

Next state table, reduced by Row elimination method, could be reduced further if equivalent states exist in the state table. Row elimination method cannot identify equivalent states. Next state table with least number of states is obtained using Implication table method.

13.5.2 Implication Table Method

Implication table method is a computational technique for state reduction. The algorithm of the method is illustrated to identify redundant states for the same hypothetical next state table in Fig. 13.20. Implication table method identifies redundant states that change to same and equivalent states.

13.5.2.1 Square Chart

Implication table method uses a square chart for filling the outcome of examining each pair of states. The square chart for next state table with seven states is shown in Fig. 13.22. The outcome of examining the states, A and B, is filled in the square, AB; the outcome of examining the states, A and C, is filled in the square, AC and so on. Filling the squares of the chart is explained.

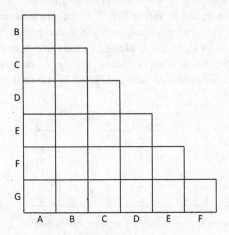

Fig. 13.22 Square chart for state table with seven states

13.5.2.2   Pairs of Non-equivalent States

Filling the squares of the chart begins with indicating the pairs of states that are not equivalent by considering the outputs when X = 0 and X = 1. A cross is used to indicate the non-equivalent pairs of states. Consider the states, A and B. The outputs of the states differ when X = 1. Hence, a cross is placed in the square, AB. The process is continued by examining A-C, A-D, A-E, A-F, A-G, B-C, B-D and so on until the states, F and G are examined. The pairs of non-equivalent states are indicated by crosses in Fig. 13.23.

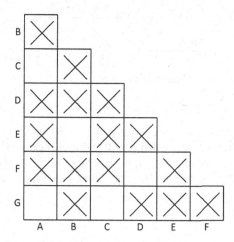

Fig. 13.23 Non-equivalent pairs of states

13.5.2.3   Implied Pairs of Equivalent States

There are five squares that are not crossed in the chart. The pairs of states, represented by the squares are possibly (implied) equivalent states as their outputs when X = 0 and X = 1 are same. The pairs of next states when X = 1 is entered below the pairs of next states when X = 0. The implied equivalence of states is verified to identify redundant states. Verification is done as explained in Sec. 13.5.2.4.

When examining the pairs of states, there could be pairs of states that change to the same states and generate same outputs. They are also entered in the squares of the chart. The pairs of same states are identified by tick mark symbols. The filling of the five squares of the chart is explained and the outcome shown in Fig. 13.24.

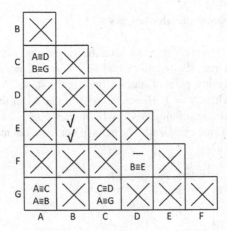

Fig. 13.24 Pairs of same and implied equivalent states

Square, AC:

The states, A and C, are examined. A remains at A when $X = 0$ and it changes to B when $X = 1$. C changes to D when $X = 0$ and it changes to G when $X = 1$. The states, A and C are equivalent if A is equivalent D when $X = 0$ and if B is equivalent to G when $X = 1$. The pairs of implied equivalent states are entered as, $A \equiv D$ and $B \equiv G$ in the square, AC.

Square, AG:

Examining the states, A and G, the pairs are implied equivalent states. They are entered as, $A \equiv C$ and $A \equiv B$ in the square, AG.

Square, BE:

Examining the states, B and E, both the states changes to the same state, B when $X = 0$ and they change to the same state, D when $X = 1$. They are shown by tick mark symbols in the square, BE.

Square, CG:

Examining the states, C and G, the pairs of implied equivalent states are entered as, $C \equiv D$ and $A \equiv G$ in the square, AG.

Square, DF:

The states, D and F are examined. D remains at D when $X = 0$ and it changes to B when $X = 1$. F remains at F when $X = 0$ and it changes to E when $X = 1$. $D \equiv F$ is a pair of self-equivalent states when $X = 0$ for the square, DF. Pairs of self-equivalent states are not entered in square chart. Hence, $D \equiv F$ is not entered in the square, DF and a dash is entered in its place. The pairs of implied equivalent states when $X = 1$ is entered as, $B \equiv E$ the square, AC.

13.5.2.4 Identifying Redundant States

The method of identifying redundant states is explained and it is shown in Fig. 13.25. The square, BE, has two tick marks and hence, the states, B and E, are directly confirmed as redundant states. State-E is eliminated. Redundant states are identified for the remaining four squares that contain pairs of implied equivalent states.

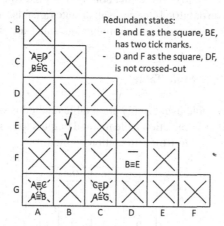

Fig. 13.25 Identifying redundant states

Square, AC:

The square contains the pairs of implied equivalent states, $A \equiv D$ and $B \equiv G$. Figure 13.24 shows a cross in the square, AD, indicating that $A \equiv D$ is false. Hence, a dashed cross is placed in the square, AC, confirming that the states, A and C are not redundant states.

Square, AG:

The square contains the pairs of implied equivalent states, $A \equiv C$ and $A \equiv B$. Figure 13.24 shows a dashed cross in the square, AB, indicating that $A \equiv B$ is false. Alternatively, Fig. 13.25 could also be used. The figure shows a dashed cross in the square, AC indicating that $A \equiv C$ is false. Hence, a dashed cross is placed in the square, AG, confirming that the states, A and G are not redundant states.

Square, CG:

The square contains the pairs of implied equivalent states, $C \equiv D$ and $A \equiv G$. Figure 13.24 shows a cross in the square, CD, indicating that $C \equiv D$ is false. Hence, a dashed cross is placed in the square, CG, confirming that the states, C and G are not redundant states.

Square, DF:

The square contains B ≡ E as the pair of implied equivalent states. Examining Fig. 13.24, the square, DF, cannot be crossed-out. It indicates that B ≡ E is true. Hence, the states, D and F are redundant states. State-F is eliminated.

Redundant states are identified after confirming that the pairs of implied equivalent states are true. Many iterations of verification are required to confirm whether the equivalent states are true or false. If the equivalent states are false, the relevant squares of chart are crossed-out. The final pairs of redundant states are shown in Fig. 13.25.

13.5.2.5   Reduced Next State Table

The reduced next state table using Implication table method is shown in Fig. 13.26. It could be observed that Implication table method outputs next state table with least number of states compared to Row elimination method.

Present	Next state		Output, Y	
state	X=0	X=1	X=0	X=1
A	A	B	0	0
B	B	D	0	1
C	D	G	0	0
D	D	B	1	1
G	C	A	0	0

Fig. 13.26 Reduced next state table using Implication table method

References

1. State Machine Design, AMD Pub. 90005, Rev. A, June 1993
2. Finite State Machines, Artix-7 10-1, Xilinx, 2015
3. Chu PP (2006) RTL hardware design using VHDL. Wiley
4. Binary Sequence Detector, AN1139, Silego Technology

Chapter 14
Technologies and General Parameters of ICs

Abstract Digital ICs are available with TTL, CMOS and BiCMOS technologies for a wide range of logic functions. The technologies and the general parameters of digital ICs are presented. Logic switching voltage levels, Noise margin, Fan-out, Absolute maximum ratings and ESD ratings for CMOS devices are some of the general parameters of ICs.

14.1 Logic Families

Digital ICs are available with TTL, CMOS and BiCMOS technologies for a wide range of logic functions. Devices are also available with ECL technology for limited logic functions. Emitter Coupled Logic (ECL) is also called Current Mode Logic (CML). A pair of bipolar junction transistors (BJTs) operates in differential mode with operating current less than the saturation current of the transistors in ECL. The unsaturated operating conditions of the transistors results in faster switching time. The obsolete logic families such as Diode Logic (DL) and Diode Transistor Logic (DTL) are not presented. Only TTL, CMOS and BiCMOS technologies are presented.

14.1.1 TTL Technology

TTL is the acronym of Transistor Transistor Logic. TTL ICs use bipolar junction transistors (BJTs) for the construction of logic gates. The data sheets of the ICs for basic gates provide the internal schematic diagrams of the gates with npn BJTs.

The power needs of TTL ICs are high. The BJTs of TTL ICs operate at currents at saturated levels. The switching speed of TTL ICs is also relatively less than ECL ICs due to minority carriers at saturated levels of current. A family of ICs is available in TTL technology with improved performance and lower power needs for various applications.

© Springer Nature Switzerland AG 2020
D. Natarajan, *Fundamentals of Digital Electronics*,
Lecture Notes in Electrical Engineering 623,
https://doi.org/10.1007/978-3-030-36196-9_14

14.1.1.1 TTL Families

Many variants in TTL families are available. ICs with operating voltages 5 V and lower are available. Manufacturer logic guides indicate the availability of ICs in TTL families with operating voltages and also provide examples of the applications of the ICs. The general logic families in TTL technology are listed below with their identification part numbers.

Standard TTL: 74XX series
Schottky TTL: 74SXX series
Low power Schottky TTL: 74LSXX series
Advanced Schottky TTL: 74ASXX series
Advanced low power Schottky TTL: 74ALSXX series
Fast Schottky TTL: 74FXX series.

14.1.1.2 TTL Families in 54 Series

ICs in the family of TTL technology are available in 54 series and they are 54XX, 54SXX, 54LSXX, etc. The devices are compatible with the equivalent ICs in 74 series. The chips of ICs in 74 series are plastic encapsulated. The chips of ICs in 54 series are hermetically sealed with ceramic encapsulation. The ICs in 54 series are used in high reliability military and aerospace applications. The datasheets of manufacturers could be referred for the availability of ICs in 54 series.

14.1.2 CMOS Technology

Complementary Metal Oxide Semiconductor (CMOS) technology uses both p-type (PMOS) and n-type (NMOS) field-effect transistors for the construction of logic gates. The datasheets of the CMOS ICs for basic gates provides the schematic diagrams with PMOS and NMOS field effect transistors. Static power dissipation in CMOS digital ICs is very low compared to TTL ICs. Static (quiescent) power dissipation is the power consumed by IC when the device is not changing its logic state. However, switching the logic gates at high frequency and with capacitive load could result in significant power consumption in CMOS ICs [1].

14.1.2.1 CMOS Families

The general logic families in CMOS technology are listed below with their identification part numbers. CMOS ICs with operating voltages 5 V and lower are available. The logic guides of IC manufacturers could be referred for additional variants in the logic families, the applications of the families, device operating voltages and the availability of ceramic encapsulated devices in 54 series. The general logic families in TTL technology are listed below with their identification part numbers.

CMOS ICs: 74CXX series and CD4000 series
High speed CMOS ICs: 74HCXX series
Advanced CMOS ICs: 74ACXX series
Advanced High speed CMOS ICS: 74AHCXX series.

14.1.3 BiCMOS Technology

The switching voltage levels (logic High and logic Low) are presented in Sect. 14.2.1 for the ICs with operating voltage, 5 V, in TTL and CMOS families. It could be observed from Fig. 14.1 that TTL and CMOS ICs are not compatible as the input and output logic voltage levels differ. TTL ICs cannot drive CMOS ICs. Passive networks or buffer ICs are required for interfacing TTL and CMOS ICs, considering the input-output current characteristics of the devices.

BiCMOS ICs have TTL-compatible inputs. TTL ICs can drive BiCMOS ICs directly without the need of interfacing circuits. BiCMOS ICs are fabricated using Bipolar Junction Transistors (BJTs), NMOS and PMOS field effect transistors. 74HCT series, 74ACT series and 74AHCT series are available and the catalogues IC manufacturers could be referred for the availability devices for various logic functions.

14.2 Generic Application Requirements

Requirements for the selection and application of digital ICs are specified in the datasheets of the ICs. The application notes of device manufacturers, journals and the proceedings of symposiums provide adequate support to design digital circuits satisfying functional and reliability requirements. The generic application requirements that are common to digital ICs for the design of circuits are presented. The requirements are:

- Logic switching voltage levels
- Noise margin
- Propagation delay
  It is explained in Sect. 3.7.1
- Fan-out
- Absolute maximum ratings
- ESD ratings for CMOS devices.

14.2.1 Logic Switching Voltage Levels

The logic switching voltage levels for 5 V Standard TTL (74XX series) and 5 V CMOS families are shown in Fig. 14.1 [2]. V_{IL} is the gate input voltage for logic Low and V_{IH} is the input voltage for logic High.

V_{OL} is the gate output voltage for logic Low and the internal logic circuit of the gate is conducting. V_{OH} is the gate output voltage for logic High and the internal logic circuit of the gate is not conducting. Logic level change operations in digital circuits are guaranteed provided:

For 5 V Standard TTL families:

- $V_{IL} \leq 0.8$ V and $V_{IH} \geq 2.0$ V
- $V_{OL} \leq 0.4$ V and $V_{OH} \geq 2.4$ V.

For 5 V CMOS families:

- $V_{IL} \leq 1.5$ V and $V_{IH} \geq 3.5$ V
- $V_{OL} \leq 0.5$ V and $V_{OH} \geq 4.44$ V.

Logic changes at other voltage levels are undefined and the undefined levels are shown in the figure. Logic guides and datasheets of manufacturers should be referred for logic switching voltage levels for other categories of ICs in TTL and CMOS families and for ICs with lower operating voltages.

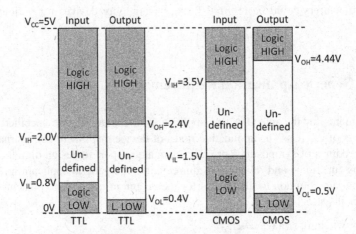

Fig. 14.1 Logic voltage levels for 5 V standard TTL and CMOS ICs

14.2.2 Noise Margin

"Noise margin" is a measure of logic circuit's resistance to undesired switching [3]. It is measured in volts. Noise Margin (NM) is defined for both logic Low and logic High. Guaranteed values for NM_{Low} and NM_{High} are obtained from the logic voltage levels for 5 V standard TTL and CMOS families shown in Fig. 14.1.

$$NM_{Low} = V_{IL} - V_{OL}$$
$$NM_{High} = V_{OH} - V_{IH}$$

Noise margins should be positive and higher values of noise margins indicate higher resistance of logic circuits to undesired switching. Noise margin for 5 V TTL families is shown in Fig. 14.2 and it is applicable for CMOS families also. It could be observed that CMOS families have higher noise margin than TTL families. Datasheets of ICs indicate that V_{OH}, V_{IH}, V_{OL} and V_{IL} and they could be used to compute the noise margins of the logic circuits of the ICs.

For 5 V TTL families:

- $NM_{Low} = 0.8V - 0.4V = 0.4V$
- $NM_{High} = 2.4V - 2.0 = 0.4V$.

For 5 V CMOS families:

- $NM_{Low} = 1.5V - 0.5V = 1V$
- $NM_{High} = 4.44V - 3.5 = 0.94V$.

Fig. 14.2 Noise margin in 5 V standard TTL families

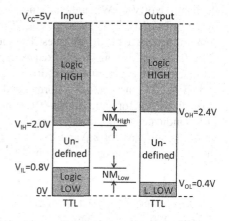

14.2.2.1 Noise Immunity

Noise immunity is relevant for digital system, having many logic circuits. It is the ability of digital system to perform the intended functions when noise signal superimposes with logic signals. Examples of noise signal sources are interference signal from power sources and cross coupling from adjacent circuits. Overall system noise immunity involves circuit related factors in addition to noise margin specifications and the general factors are line impedances, circuit output impedances, and propagation delay [3]. Noise sources and the methods of improving noise immunity are explained in the reference [4].

14.2.3 Fan-Out

The output of a logic gate is either High or Low and it drives another logic gate in digital circuits. Fan-out is the maximum number of logic gates that could be connected to the output of the driving gate without degrading its performance. The method of computing the fan-out for TTL and CMOS families is presented.

14.2.3.1 Fan-Out for TTL Families

When the driving gate conducts, the output logic state of the gate is Low, maintaining V_{OL}. The driving gate is also sinking currents from the driven gates i.e. the currents from the driven gates also flow through the driving gate. The sinking currents cause the increase of V_{OL}. When the driving gate is non-conducting, the output logic state of the gate is High, maintaining V_{OH}. The power supply (V_{CC}) of the driving gate sources currents to the driven gates. The sourcing currents cause the decrease of V_{OH}. If the fan-out of driving gate is exceeded, changes in output switching voltage levels degrade the noise margins of the driving gate.

Fan-out is computed using normalized Unit Loads (U.L.) for TTL families. The normalized U.L. for logic High is $40\,\mu A$ and it is $1.6\,mA$ for logic Low [5]. Datasheets of ICs indicate output current High (I_{OL}) and output current Low (I_{OH}). Fan-out for an IC is computed for I_{OL} and I_{OH} using the normalized U.L. The smaller value of the two computations is the fan-out of the IC.

Example-1: IC 74150

$I_{OL} = 16\,mA$; Fan - out $_{Low} = I_{OL}$ (Normalized U.L. for logic Low) $= 16/1.6 = 10$

$I_{OH} = 800\,\mu A$; Fan - out $_{High} = I_{OH}/$(Normalized U.L. for logic High) $= 800/40 = 20$

Fan - out for 74150 $\leq$ 10 Unit Loads.

Example-2: IC 74ALS138

$I_{OL} = 8 \, \text{mA}$; Fan - out$_{Low} = I_{OL}/(\text{Normalized U.L. for logic Low}) = 8/1.6 = 5$

$I_{OH} = 400 \, \mu\text{A}$; Fan - out$_{High} = I_{OH}/(\text{Normalized U.L. for logic High}) = 400/40 = 10$

Fan - out for $74138 \leq 5$ Unit Loads.

14.2.3.2 Fan-Out for CMOS Families

Input capacitance (C_{IN}) and output load capacitance (C_L) decide the fan-out of driving gate for CMOS families. C_L is denoted as C_{PD} in the datasheets of ICs as it decides the dynamic power dissipation also for CMOS ICs. Excessive fan-out degrades rise and fall times of driving gate in CMOS families. Fan-out for CMOS families is given by the ratio of C_{IN} and C_{PD} [6].

$$Fan-out = \frac{C_{PD}}{C_{IN}} \text{max}$$

14.2.4 Absolute Maximum Ratings

Datasheets of ICs specify absolute maximum ratings for electrical and thermal parameters. Electrical parameters are voltages and currents. Thermal parameters are storage and junction temperatures. The parameters are the set of reliability related stress factors. Thermal resistance is also specified in the datasheets of ICs. Exceeding the maximum values of the stress factors degrades functional performance and reduces the reliability of ICs.

Operating junction temperature of ICs is computed using thermal resistance, power dissipation and operating ambient temperature. Computing power dissipation for TTL families is straight forward using V_{CC} and I_{CC}. Static and dynamic power dissipation using C_{PD} should be computed for CMOS families [1]. Reference [7] provides additional application information on absolute maximum ratings including overstress analysis for electronic designs.

14.2.5 ESD Requirements for CMOS Devices

When certain types of materials are rubbed against each other, static electrostatic charge builds-up on the materials. One of the materials is positively charged and the other is negatively charged. The charges remain on the materials even after they are separated. The level of the charges depends on the type of materials and ambient relative humidity. The electrostatic charges in the material are discharged when the material is grounded. The discharge current is usually high for short duration. The event is called Electrostatic Discharge (ESD).

14.2.5.1 Primary Sources of Electrostatic Charges

ICs in CMOS families are highly susceptible to ESD induced damages although low power TTL devices are also susceptible to damages. There are two primary ways by which CMOS devices could be damaged. Human body gets positively charged in many ways such as while walking on a carpet or an operator picking polythene bags from assembly tables. Charges are transferred to the ICs by touching the pins of ICs with bare hands. When the ICs with electrostatic charges are grounded or mounted in circuits, electrostatic discharge occurs through the ICs destroying the devices. ICs also might have acquired electrostatic charges from other sources during handling and the ICs are destroyed when grounded or mounted in circuits.

14.2.5.2 ESD Ratings and Requirements

JEDEC (Joint Electron Device Engineering Council) standard specifies Human Body Model (HBM) and Charged Device Model (CDM) for qualifying integrated circuits for withstanding capability against ESD damages. Device manufacturers specify ESD ratings (V_{ESD}) in the datasheets of ICs for HBM and CDM.

Although ICs have internal protection circuits to prevent ESD damages, it is necessary to adhere to storage and handling instructions for CMOS ICs. Ensuring relative humidity for ambient at 40–60%, storing CMOS devices in magazine cases or conducting trays and wearing appropriately grounded wrist strap directly on the skin are some of the requirements. Guides [8] of manufacturers could be referred for additional information for preventing ESD damage to integrated circuits.

14.3 Logic Pulser and Probe

Trouble shooting is part of design process. In analog circuits, a small signal of appropriate frequency is injected and the signal is monitored for performance at various points to identify faulty analog circuits. Logic pulser and probe is a hand-held tool and it is used for identifying faulty digital circuits without de-soldering components from printed circuit boards. The tool could be used either as a pulser or as a probe.

Logic pulser and probe has a single-shot, high current pulse generator. It can source sufficient current to force low outputs out of saturation to a high state and sink sufficient current to pull normally high outputs below the logic zero threshold [9]. The probe is capable of detecting logic High, logic Low and voltages in the undefined region. The detected output is indicated by color displays. Facility exists in the tool for trouble shooting TTL and CMOS logic circuits. Logic pulser and probe has high input impedance and it does not load the performance of digital circuits. It is also used for servicing digital circuits.

References

1. CMOS Power Consumption and C_{pd} Calculation, SCAA035B, June 1997, Texas Instruments
2. Logic Guide (2017) Texas Instruments
3. General Information for MECL 10H^{TM} and MECL 10K^{TM}, Application Note TND309D, June 2002, ON Semiconductor®
4. Elgamel M, Bayoumi M (2006) Interconnect noise optimization in nanometer technologies. Springer, Heidelberg
5. LS TTL Data, DL121/D, Rev. 6, ON Semiconductor®, Jan 2000
6. Bhushan M, Ketchen MB (2015) CMOS test and evaluation. Springer, Heidelberg
7. Natarajan D (2015) Reliable design of electronic equipment: an engineering guide. Springer, Heidelberg
8. Guide to Prevent Damage for Semiconductor Devices by Electrostatic Discharge (ESD), ESD prevention, C11892EJ1V1IF00, NEC Corporation, Jul 1997
9. Adler R, Hofland JR (1972) Logic pulser and probe: a new digital trouble shooting team. HP J

Appendix
Erasable PROMs

A.1 Erasable PROMs

Erasable PROMs (EPROMs) are storage devices and they are capable of storing larger data with fast write, read and erase features. Programmed codes in the devices remain permanent until they are erased. Erasing operation restores the devices to their original (erased) state for programming new set of codes. New set of codes could be programmed in the devices after erasing. EPROM has an array of MOSFET memory cells. The structure and logic levels of MOSFET memory cell are presented.

A.2 Structure of Memory Cell

The structure of a memory cell is similar to that of a general purpose MOSFET (Metal Oxide Semiconductor Field Effect Transistor). P-substrate is used for the memory cell. Apart from source, drain and control gate connections, the memory cell has a floating gate, which is placed in between the control gate and the P-substrate. The representation of general purpose MOSFET device and the floating gate MOSFET memory cell with P-Substrate are shown in Fig. A.1. EPROMs contain an array of floating gate MOSFET memory cells. They are also called floating gate devices.

A.3 Logic Levels

The logic levels of memory cells are similar to logic levels of ROM devices. Initially, all the floating gate MOSFET memory cells of erasable PROMs are in conducting state. The conducting state of the memory cells is defined as logic level, 1 and the memory cells are in erased state. New EPROMs are delivered in erased state. It is

© Springer Nature Switzerland AG 2020
D. Natarajan, *Fundamentals of Digital Electronics*,
Lecture Notes in Electrical Engineering 623,
https://doi.org/10.1007/978-3-030-36196-9

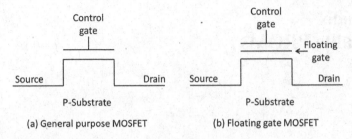

Fig. A.1 Representation of floating gate MOSFET memory cell

equivalent to having fuse-diode circuits for all the logic paths in PROM and the conducting logic paths are at logic level, 1.

The non-conducting state of floating gate MOSFET memory cells is defined as logic level, 0. Selected memory cells are programmed to the logic level, 0, as per code requirements of logic circuits. It is equivalent to blowing the fuses of selected logic paths in PROM for programming.

A.3.1 Electrically Programming Floating Gate Memory Cells

Assume that EPROM is in erased state initially. If normal operating voltages are applied to the drain and control gate terminals of the MOSFET memory cells, the memory cells would conduct. All the MOSFET memory cells of EPROM are at logic level, 1.

The control gate voltage of selected MOSFET memory cells is increased to higher than the normal operating voltage (approximately 20 V) for programming the cells to logic level, 0. The electrons (charges) are trapped in the floating gate of the cell, pushing the memory cells into non-conducting state. The selected memory cells are programmed to logic 0. Normal operating voltage is applied to memory device for reading the programmed codes (ones and zeroes) in EPROM. As the voltage for reading the codes is less than the voltage for programming the memory cells to logic level, 0, the programmed codes are not altered. The programmed codes of floating gate MOSFET memory cells remain until they are erased.

A.3.2 Erasing Operation

The charges on the floating gate are neutralized in erasing operation. Two types of devices are available with two methods of erasing. They are Electrically Erasable PROM (EEPROM) and Ultra-Violet Erasable PROM (UVEPROM). The operation of the devices is presented.

A.4 UVEPROM

Programming the memory cells of UVEPROM is done electrically. Erasing operation is performed using ultra-violet rays. UVEPROM has a quartz window for erasing operation. The chip of the device is exposed to ultra-violet rays through the window. The charges on the floating gates of programmed cells are neutralized and the cells are restored to the conducting state. The logic levels of the memory cells becomes 1. Generally, UVEPROM is exposed to UV light through the window prior to programming.

A.5 EEPROM

Electrically Erasable PROM (EEPROM) are programed and erased electrically. Erasing is performed by applying approximately -20 V to the control gates of the memory cells for discharging the trapped electrons on the floating gates into the P-Substrate of EEPROM.

A.6 Flash Memory

Flash EEPROM devices are commercially known as flash memories. The devices are characterized by fast programming and erasing capabilities. The devices are programed and erased electrically. Programming is performed either by using the avalanche breakdown method or Fowler-Nordheim tunneling method as per device manufacturers. Erasing operation is performed by the tunneling method. Programming and erasing operations could be performed under in-circuit conditions. Mass storage is one of the popular applications of flash memory devices in computers. Flash memory devices with higher than 10 GB capacity is available from manufacturers.

Index

© Springer Nature Switzerland AG 2020
D. Natarajan, *Fundamentals of Digital Electronics*,
Lecture Notes in Electrical Engineering 623,
https://doi.org/10.1007/978-3-030-36196-9

Printed in the United States
by Baker & Taylor Publisher Services